HAIR MAT

HAIR MATTERS

Joshua and Daniel Galvin

© Joshua and Daniel Galvin 1985

All rights reserved. No part of this publication
may be reproduced or transmitted, in any form
or by any means, without permission

First published 1985 by
Higher and Further Education Division
MACMILLAN PUBLISHERS LTD
London and Basingstoke
Companies and representatives
throughout the world

Printed in Hong Kong

British Library Cataloguing in Publication Data
Galvin, Joshua
Hair matters
1. Hair – Care and hygiene   2. Beauty, Personal
I. Title   II. Galvin, Daniel
646.7'08842   RL91
ISBN 0-333-26541-6

To Mum and Dad with all our love

Acknowledgements

We would like to thank everyone who helped make this book possible.

Our sincere thanks to our staffs both past and present for their contribution to our success.

Grateful thanks to Wella (Great Britain) Limited for allowing us to use their graphic illustrations in this book. The cover illustration is by courtesy of *Woman Magazine*.

Particular thanks to the top make-up artists and photographers we have worked with throughout the years. Also to the press, radio, television and the film industry for their support through the years.

Our very special thanks to Eileen McCarroll, former editor of *Hair and Beauty* and *Hairflair* for her patience, perseverance and skilful interpretation of our thoughts on to paper for you.

To all our Clientele we are always at your service.
      Sincerely
           Joshua & Daniel

Contents

Preface   x

Introduction   xi

1 The Inside Story   1
    From the outside looking in   2
    Inside information   3
    A few hairy facts   6

2 Know Your Type!   9
    What class are you in?   10
    Growing pains   13

3 Going for a Clean Image   21
    All about shampoos   21
    Shampoo dos and don'ts   23
    How to wash hair   24
    Hair washing questions and answers   27

4 Be a Shining Example   31
    Conditioners on the market?   31
    Get into condition   36

Contents

5 Good Condition – More Than Just Conditioner   39
    The importance of diet   39
    How to use your tools correctly   42
    Holiday hair care   47

6 The Secrets of a Successful Salon Visit   51
    Finding the right salon   51
    Make a friend of your hairdresser   54

7 Suit Yourself!   59
    Styles to suit body and face shapes   63
    Tackling problem areas   68
    The importance of a good cut   71
    Styling tips   74

8 Get Set   75
    Setting techniques   75
    Setting agents   81

9 When the Heat Is On   83
    Drying techniques   83
    Styling pointers   90

10 A Permanent Alternative   93
    Perming techniques   96
    Professional versus home perming   101
    The pros and cons of perming   103
    A permanent solution   105

11 Colour Is What Counts   107
    Temporary rinses   109
    Semi-permanent rinses   110
    Tints   111
    Bleaching   115
    Lit up in lights   116
    Vegetable colours   120
    Colours of the spectrum   124
    Powder sprays   125

Contents

12 Child's Play   129
    The early years   129
    Teen times   133

13 On Longer Lines   135
    Conditioning care   135
    Styling options   138
    To cut or not?   139

14 Dressing Up   141
    Bridal beauty   141
    Hair accessories and hats   143
    Fantasy styling   144

  Glossary   145

  Index   148

Preface

The exciting world of hairdressing is one we both love and thrive in. We have never wanted to be anything other than good hairdressers, and have worked long and hard to achieve success in our respective areas of expertise. But what we have learned means nothing if we do not pass on some of the experience we have gained in forty years in hairdressing. We have, of course, shared our knowledge with colleagues at hairdressing shows, seminars and teach-ins, but we have always been a little sad on these occasions that we were not also reaching the people that matter most, the clients who make our jobs so enjoyable and worth while. We are glad to say this book now remedies that situation. All that is left for us to hope is that you will follow our advice to make the most of your hair and that you will enjoy yourself while doing so.

JOSHUA GALVIN
DANIEL GALVIN

Introduction

Many people are actually afraid of their own hair. This may seem hard to believe and rather an exaggeration but a lot of women don't know exactly how far they can go with their hair and so give up on it altogether. True, hair does have limitations and you cannot always get what you want from it, but there are alternative approaches. We tell you in this book what some of those are, and hope by our explanations to increase your confidence.

The big thing about hair is that it has progressed from being merely a covering for our scalps to being a positive beauty asset that enhances our natural features *if we let it*. Hair is the only part of the body with such versatility – you can completely change your looks with the hairstyle you choose. But hair has to be looked after, treasured as a cashmere sweater might be, so that it can be moulded to your requirements. Unfortunately, too many clients and even some hairdressers abuse hair horribly, often quite unintentionally. The purpose of this book is to explain the structure of hair, how to care for it, what perming does to hair and why certain products are better for the hair than others.

We want to make the public far more conscious of caring for hair at home, but only to supplement the services received at the salon, not to replace them. In hairdressing today there are

Introduction

many advanced techniques that should not be in the hands of non-professionals and we don't feel it would do you any service to tell you all about them. Instead, we have produced this book as a guide to how you can best use your hairdresser and the professional services available *and* keep up the necessary maintenance at home. We want to ensure that your hair is in better shape and condition after you have read this book than when you started it, an improvement achieved, we hope in friendly consultation with your stylist, and with you better informed about what's going on. Here's to healthy hair!

1 *The Inside Story*

What is hair made of? How do I know my type? Is there a special way to wash or dry hair? Am I using the right kind of products? Does what I eat really make any difference? What style would best suit my features? All of these questions, plus many more, are the sort you should be asking and getting answered if you really intend to give your hair the care and attention it requires for healthy growth.

Yet our guess is that very few of you, much like the population as a whole, have considered hair care in quite such a serious light before. Correct? Well don't feel too bad because just being honest about this means half the battle is already won, and you *are* reading our book which is definitely another step in the right direction. Nor do we believe you are completely to blame for having this basic lack of interest in your hair. In fact, you will probably be surprised to hear us say that hair itself is mainly at fault. By being such a resilient, pliable material, we are led into taking our own too much for granted and to some extent, almost forgetting it is even there!

From an early age hair care comes a long way down on most people's list of personal grooming priorities. Usually washing, drying, brushing and styling, when really essential, tend to be the sum and substance of the average weekly routine. Only when something goes drastically wrong do people appreciate that looking after hair is far more involved than they had previously imagined. Perhaps their new style won't settle down or an expensive perm falls out too quickly. Maybe dandruff has appeared on their scalp for the first time ever, making them embarrassed and self-conscious. Whatever the problem is, when they find their hair no longer performing on cue just as an important date or interview is on the cards, they rush round to the local salon expecting everything to be put right on the spot.

Now we are not saying hairdressers cannot carry out minor miracles, because they can and do, but they must have something reasonable to work with. If your hair is playing up because of months of neglect then obviously it will take as many months again to recover. To expect an instantaneous cure is as ridiculous as going along to your doctor with a broken leg and telling him to mend it there and then.

Some clients cannot see this. They don't understand that hair care means *constant* care and not just when they feel in the mood. The old saying, 'You only get out of something what you put into it' is never truer than when applied to hair. Unless you consistently invest time and effort to look after the raw material, you are not being fair expecting wonders from your hair or your hairdresser. But to handle hair successfully requires some 'inside' information, so READ ON.

From the outside looking in

If you could travel all the way down the length of a solitary strand of hair to below scalp level, you would see how the hair emerges from a narrow, slanting passageway under the skin. This small indentation or pocket in the skin is called a *follicle*. At its base lies a small nodule, the *papilla*, which is connected to an ingenious network of blood vessels to receive the oxygen and food vital for hair growth. Think of a car factory production line, and of the conveyor belt carrying the parts that make the car, and you'll have some idea of how the papilla receives the components necessary to maintain hair growth from the blood supply. If this supply is poor, growth is too, and in severe cases hair loss occurs. Scalp massage is frequently recommended by trichologists (hair and scalp specialists) as an aid to improving the blood's circulation and thus stimulating growth.

During the growing period hair is full of life but as the soft cells are pushed up their passageways into daylight by other growing cells, they die and harden. The dead cells are the end result we see on our scalps, and for that matter all over our bodies, as hair. So if anyone ever tells you your hair is looking *dead*, don't argue because technically the observation is quite accurate, though in all likelihood your 'friend' is actually referring to your hair's condition. The fact of the matter is this: the hair you can see is so dead that when you get a haircut you don't feel a thing. When pulled or tugged though, that's quite a different story. Nerve endings linked to the hair buds transmit a warning to the brain that if the pulling doesn't soon stop, the hair could be badly damaged.

About 97 per cent of hair is made up of the same substance as your nails, a protein called keratin; plus certain trace elements that are also present in the body. The main constituents of all proteins are amino acids, which for hair are transformed into keratin by a particular group of cells in the papilla. The method of their construction can be likened to the way a house is built with bricks.

Some amino acids are linked in long chains to form what we call the *cortex*. These chains are bound together, by the sulphur and hydrogen bonds that are also part of hair, to create very long units which make hair the hardy structure it is. The 'hardening' process that takes place somewhat resembles the way liquid rubber is vulcanised to make things like car tyres, for example, more durable.

Inside information

When you eat an orange you have to remove the skin and break through the pith to reach the flesh of the fruit. If we could dissect a hair strand in much the same way and then examine the result under a microscope, three similarly distinct

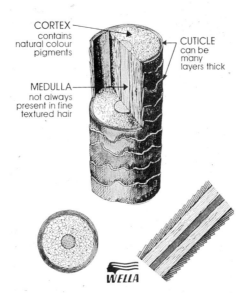

Cross-section of hair showing the cuticle, cortex and medulla

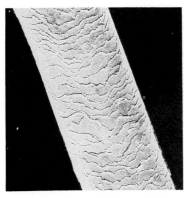

A healthy strand of hair with the cuticle closed

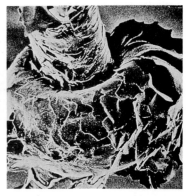

A damaged hair strand where the whole of the cuticle has opened up like a cabbage

The Inside Story

layers would become visible to us – the *cuticle*, the *medulla* and the *cortex*. The cuticle lies on the outside, the medulla makes up the central core and the cortex is sandwiched between the other two.

The *cuticle* has flat, overlapping, transparent scales, a bit like snakeskin to look at, which form a hard, 'keratinised' protective shell around the other layers. The free ends of these scales point towards the tip of the hair shaft away from the root. This orderly formation means any dirt particles that find their way into the system can easily be removed by brushing down from the base of the scalp to the hair tips. When hair has a super shine these scales are lying flat and light is reflected off them. Conversely, when the scales flap up or split, as happens when hair is extremely dry, has been damaged by chemicals, or has suffered physical abuse, hair looks generally dull and lifeless.

The *medulla* is at the centre of the hair shaft and in some cases runs the entire length of the hair whilst in others it is broken at intervals or missing altogether (usually from very fine hair or from the hair of older people). The true purpose of the medulla is still unknown and its presence or absence seems to make little difference to styling results or to the effectiveness of chemical treatments.

The bulk of your own hair's particular characteristics is determined by the *cortex*'s fibre-like cells. Between 75 per cent and 90 per cent of hair is cortex and its make-up of inherent properties affects the strength, texture, elasticity, length and quality of your hair and to a certain extent decides whether it is curly or straight. (These cells are also the ones that bleaches, dyes and perms have to penetrate to become active.) Another point worth noting here is that constant perming, colouring or bleaching, or the application of one of these treatments on top of the other, without a breathing space in between, damages these cortex cells so that the hair eventually becomes weak, brittle and breaks easily.

The cortex cells also contain the pigments that give hair its natural colour. Pigment granules are deposited in the cortex as it grows from the papilla. In most cases they contain over 90 per cent pure *melanin* and can be large or small, incorporate varying amounts of pigment, be gathered in clusters or spread evenly throughout the cortex. If no pigment is produced at this stage the hair is left totally colourless.

Grey hair is a combination of white hair and the hair's own remaining natural colour and appears with advancing years when pigment production begins to slow down. Completely *white hair* is altogether harder to explain, but some people think severe shock plays a major contributory role in its appearance. *Dark brown* hair is mainly black and brown pigments while *ash blonde* has predominantly yellow pigment. With *redheads*, the lighter varieties contain a mix of red and yellow pigments while the darker shades are a combination of red and black pigments.

Strong natural colour obviously goes a long way in making the difference between hair looking vibrant, or mousy and dull. But to look healthy and clean, all hair must have sheen. This sheen, plus softness, is supplied to your hair by its very own conditioner, *sebum*, a natural oil composed of waxes and fats which is provided by the *sebaceous glands*. Some of the glands open directly on to the scalp but the majority are linked to the hair follicles and release sebum into them. As a lubricant sebum gives an excellent protective coating to the entire hair shaft, keeps the cuticle scales lying flat (and shiny), and helps the hair to retain natural moisture. Occasionally though, the end result can be more a case of the plain old greasies because too much sebum has been released at one time. The amount produced varies with the individual and is controlled by hormonal activity, which in turn determines whether your hair is normal, dry or greasy.

Factors affecting the degree of sebum secretion include diet, age, health, body changes and nationality. For example, an adolescent boy often complains of having terribly greasy hair; this is because his glands are working overtime at this point in his life due to the normal hormonal imbalances at puberty.

A few hairy facts

Most hair grows about half an inch per month. Growth is faster in women than in men and is greater in summer (when encouraged along by the sun's rays) than in winter. The longer hair is, the older the ends, which is why a short style keeps in shape better. On average, every head has between 120,000 and 150,000 hairs (though this can vary with different colours; redheads normally have the fewest hairs – about 90,000). Each head loses between 50 and 100 strands daily so don't worry if you see loose hairs on your comb or brush, or in the bath. As long as your hair is not coming out in clumps this moulting process is perfectly natural. Should you ever lose large clumps, however, visit your doctor for advice because there is a strong possibility that something is seriously wrong at scalp level which might require expert trichological treatment.

Growth rate varies on different areas of the head and this is the reason a cut can appear to go out of shape very quickly. There's more hair at the back of the head and on the crown than anywhere else, and the angle at which it grows or lies most comfortably is dependent upon the direction in which the follicles are pointing under the scalp. This is not the same for any two people but as a general rule, crown hair is more 'stand-uppish' than the sides. Hair never stops growing – it simply slows down with age.

Hair also goes through what might be termed a 'rejuvenation' process, falling out and being replaced in an overlapping sequence so that a normal healthy scalp has roughly the same number of hairs all the time. What happens is that growth occurs in cycles with every single follicle going through both growing and resting periods. An average hair has a lifespan of six to seven years. The 'rest' period can be as little as a few weeks or as long as two to six months and occurs when the follicle shrinks. At the rejuvenation stage hair loosens from the papilla but remains in the follicle until brushed or shampooed out or until the production of the new hair begins and the old one is pushed out and replaced by the new hair a couple of months later. At any one time, over 90 per cent of the scalp's follicles are at different stages of development whilst the remainder are resting before falling out. The growing phase is longest for women between the ages of fifteen and forty due to an increase of the female hormone, oestrogen, during these years. Interestingly, after the female menopause, oestrogen production decreases and a woman's scalp hair becomes very much like a man's exhibiting similar signs of thinning patches.

As we've said already, the direction in which your hair grows depends upon the angle at which each follicle lies under the scalp. Thus a natural parting and any irregularity, like a cowlick, is produced by follicle groups positioned at specific angles. But you may be surprised to know that these angles can be altered in special circumstances. Ever had a fright and gone goose-pimply all over and felt as though the hairs on your body were standing on end? Yes? Well, they were! When your nervous system signals to the brain that you have literally gone 'cold with fright', the sudden cooling of your body causes an amazing little muscle at the follicle base, the *arrector pili*, to contract. This movement pulls the hair sharply into an upright position, where it remains until the crisis is over.

Another interesting fact about hair is that when untreated, that is, when it has never been chemically processed by perming or colouring, it is very elastic and will stretch as much as half its normal length. The 'stretch' is even greater in water but the degree of elasticity definitely depends upon the hair's overall condition. Healthy wet hair returns to its original length intact when dry, but damaged hair loses elasticity at a dramatic rate and contraction doesn't come easily, if at all. Very dry air dehydrates hair and causes slight temporary shrinkage; the same hair later immersed in water will soak up and absorb half its own weight of liquid – the hair swells and becomes longer and curlier. Without this elastic quality, many of today's hairdressing techniques could not be employed.

2 *Know Your Type!*

What a funny lot we all are. Half of us seem to spend our lives wanting what the other half has and ironically, they are probably envious of what we have! Of course, there's no area in which this is truer than with natural hair types. Given a chance, you can be sure many of your natural 'curlers' would love to go 'straight' and what wouldn't those 'thickies' give to be 'less coarse'.

To trade in the old model for a new one is a desire that probably comes about because many people feel totally unable to cope with their hair as it is. Desperation turns to self-pity and the sincere belief that if their hair were only short/long; brown/black; thick/thin . . . they would look so much better. This whole attitude is wrong even if there is some truth in the idea. Never think, 'There is absolutely nothing I can do to improve my hair. I don't care what anyone else says.' Everybody can look great *now* with a little bit of help, a small amount of effort and a lot of confidence. The thing to remember is, you must understand your hair and the way *your* model behaves before you will achieve anything. So first things first, get acquainted with *your* type.

What will be possible, stylewise, will depend entirely on the hair's *texture* and *quality*. Choose a style only after you have given some thought to these two factors and then realistically appraised what results you can reasonably expect. If you still feel all at sea once you know the facts, seek your hairdresser's advice. All hairdressers love a challenge! He or she will create a look which is right for your hair and give you tips on home hair care between salon visits so that you have a round-the-clock routine for making the most of yourself.

What class are you in?

Hair varies with the individual much as skin type does. The various groups are genetically acquired, controlled by hormones and have different textures relative to age and racial characteristics.

On a general level, three distinct racial groupings of hair types can be identified:

Negroid races – hair is dark, curly and is flat when seen in cross-section.

Oriental races – hair is dark, straight and is circular in cross-section.

Caucasian races – hair is light, wavy and is oval-shaped in cross-section.

A single head of hair may include several textures. Coarse sections are quite usual on the nape area while the temples and the front hairline have very fine strands. Hair texture is, if you recall, decided upon at root level in the cortex, and the follicle shape determines whether it is curly or straight, thick or thin. Nothing can permanently change this shape, not even chemicals, although on rare, inexplicable occasions hair has been known to alter from curly to straight completely of its own accord.

Hair colour can also give some indication of texture type. Blonde hair is often the finest, red the coarsest and brown somewhere in between. There are four main texture categories – normal, fine, medium and coarse.

Know Your Type!

Normal hair

Hair of this texture is neither too thin nor too thick, is smooth, strong, elastic and rich in colour. Normal textures are not very common and so hair tends to fall into slightly confusing sub-sections of the other three. For example, you find some Oriental people with coarse hair that is thin while some Scandinavians have fine hair which is thick. The different textures of straight and curly hair are listed below but bear in mind that there is no truly straight hair – all hair contains a certain slight wave movement. Discover what class your hair is in, remembering that from time to time this can alter marginally as other body changes occur.

Fine hair

To work out if your hair comes from a 'fine' family or not is relatively simple. Fine hair is very soft, light and likely to be 'flyaway'. If you recognise these symptoms from your own hair's behaviour then from bitter experience you already know just what careful handling it demands. This texture is fragile and cannot be treated nearly as roughly as a stronger type might let you get away with. Damage can be done almost by accident and because of the hair's poor quality, design lines fall out quickly. The temptation to overwork this hair is high for this very reason. To maintain a particular look the hair is frequently subjected to over-use of rollers and dryers, and becomes a greater mess as a consequence. People with fine hair generally have to visit a salon regularly because growth is poor, slow and the ends break easily. A cut or trim is an absolute must every few months if any sort of styling is to succeed. To help fight styling difficulties use a shampoo with 'body' and a cream conditioner to harness the flyaway habit. Another handy aid occasionally is hairspray, and a perm or colour will add bulk.

Fine and thin The black sheep of the 'fine' family, this type really creates chaos and there's nothing you can do. The problem here is essentially one of less hair growth to the square inch than is normal, with each hair having a very thin cortex layer into the bargain. Because the scalp has fewer follicles capable of hair production there is an actual lack of hair quantity and this situation is irreversible. No offence intended if this is your own type but of all the hair categories, this one is the worst of the lot. Sadly, there is not a great deal we can do for it. Don't expect to see a beautiful style in a magazine and be able to go along to your stylist for a duplication on you. Accept what you've been born with; face facts and trust in your hairdresser's skill to create a look that takes all the disadvantages into account and leaves you with a sense of being in control.

Fine, limp and straight   Hair that's flyaway, lacks body and droops all over the place is the family 'flop'. A set can help by giving extra body but will not hold for any length of time, though hairspray will temporarily prolong its lifespan. Even a perm tends to flop quickly in fine hair because as it grows out, there is no body or volume left at the roots. While a curl effect may still be evident on the ends, the whole style falls flat because of insufficient body close to the scalp. Unfortunately, this hair often lacks lustre and looks limp too.

Probably one of the quickest ways of damaging the 'fine flop' is to opt for too much chemical processing. Understandably, perming and colouring hold strong attractions for a client at her wit's end with hair that is so irritatingly uncontrollable. But a 'Catch 22' situation soon arises and the client is worse off than when she started. Another pain in the neck for this unruly mop is static electricity. This is where a suitable conditioner comes into its own, smoothing matters out as it were.

Fine and thick   The notion that hair can be both fine and thick might appear a strange one but such a texture does exist and is particularly prevalent in people of the northern Scandinavian countries. Thick, straight and baby fine, this texture is another that needs constant mothering. The sheer abundance of hair gives an impression of weight and strength, disguising the true delicate nature of every strand. Those who have this texture sometimes erroneously believe that a lot of hair is synonymous with strength and consequently, due to heavy handling, most split ends are found in this group. A slightly elastic quality in the hair means that if the top layers are handled extensively during setting, drying or brushing, splitting occurs and shows rapidly even though the hair is essentially thick. However, a perm put into this type does have staying power and is a good styling alternative.

Fine and awkward By awkward we mean with a slight positive movement of its own, though not necessarily natural curl; simply, the hair moves in a soft way on some parts of the head and not on others. No matter what you do, a few drops of rain or beads of sweat can destroy a style in a moment, with the hair falling much where it pleases. Sometimes life is further complicated by other features – like a double crown. This phenomenon is most common in awkward heads and the scalp does indeed have two crowns – one in the usual place and another just below. Strands of hair can be seen sticking out in all the wrong places. However, clever hairdressing or setting works wonders and can produce an effective camouflage. A perm will make the stray hairs disappear altogether but, regrettably, only until the curl grows out. One-length styles are also an excellent option.

The problem may be hair growing high on the head creating a high hairline at the back of the head, in front of the ears or on the forehead. The answer to these is to employ the hair's own weight and length to fool everybody else into thinking you have more hair than you really do. For styling suggestions, speak to your hairdresser.

Fine and curly In our opinion, this hair is the most beautiful of the fine textures, even when of the thin variety. With a good cut it can look fantastic. The main thing to remember is to work with the hair in the direction it grows or you'll have a permanent fight on your hands.

All fine heads of hair have a number one enemy – grease.

Regardless of styling, grease has to be taken into consideration and the more you brush and fiddle with your hair, as you'll be inclined to do, the worse the situation will become. The secretion of grease is unusually high because, as a general rule, there are more hairs on a head of fine hair, and therefore more follicles, than on any other type. Every manipulation of the hair stimulates the oil glands and you won't ever totally control the flow. The best you can hope for is to bring the amount of sebum down by reducing washing, brushing and fiddling sessions. If grease is virtually dripping down your hair, embarrassing and upsetting you, wearing a wig may be the only answer until the problem is tackled and overcome.

Medium hair

Here we are mainly talking about mid-European type hair – neither too fine nor too coarse, and easy to manage. Generally, coloration is of a medium shade too – mousy; or if blonde, of a darker tone; or a warm chestnut. This hair lends itself especially well to all types of colouring techniques. Medium textures are flexible and suitable for most hair work but conversely, if tonged, rollered or lacquered to excess, the hair goes out of condition fast. Prevention is definitely preferable to cure.

Coarse hair

This strong hair type is prevalent in the Oriental and American Indian races. In fact, most dark hair shades are coarse textured with the exception of the Negroid, which has distinctions special to itself. The Indian and Middle Eastern populations have strong hair and the farther you travel east, the coarser the hair becomes – the Chinese and Japanese peoples have an even stronger variety.

Oriental type hair is straight like a cable, circular in cross-section and has a thicker than average cuticle and cortex. A perm or set takes rapidly but is then almost as quickly pushed out. Obviously a perm is not visibly pushed out – it has to grow out – but a demarcation line is very evident quite soon afterwards. The thicker cuticle and cortex also make this texture less suitable for colouring, lightening and waving because the products have a harder time penetrating the additional mass.

Thicker Oriental hair is, strangely enough, often accompanied by a thin and irregular hairline around the face and neck. Certain problems that present themselves in the hairline area can only be overcome by covering them up, and this is such a one. If exposed, particularly where hair only grows to a certain length, the fault shows up as a small growth patch. Crown growth is often strong in coarse hair, and wearing a short style places limitations on variability, so stick to longer looks.

Slightly coarse/lush Basically a beautiful, thick head of hair found mainly in Mediterranean peoples, the populations of the South Sea Islands, the northern Japanese islands and Scandinavia. The hair waves and curls, swings and bounces superbly; it is abundant in growth and can stand a lot of rough punishment. Amazingly, many women with this hair type hate it. There seems to be a certain monotony in this hair type's predictable good behaviour that gives rise to a different kind of frustration. However, countless models bless their hair if it is this texture for the styling versatility it allows them. Given a good cut, lush hair can be shampooed and left to dry or if long, taken up into a pretty chignon shape. After a swim, one good shake brings this hair back into place. Lucky is the client with lush hair, and money spent on care and maintenance is well spent.

Frizzy and thick Found mostly among Latin races and North African peoples, frizzy hair is thick hair and is quite a problem: grown long, the frizz curls; left short, the frizz is to all intents and purposes completely uncontrollable. Heat and humidity attack the cuticle layers making them swell and stick up. Setting has a slight straightening influence for a while but the minute the atmosphere is damp, it's back to square one.

Negroid hair plaited in classical corn-rows

Chinese hair stack permed

A full, feminine blow-dry style on Latin-type hair

Fine Northern European hair cut into a sleek bob

Know Your Type!

Chemical straightening is a possible solution but neither of us likes this process much because it places a lot of strain on the hair. We'd much rather see people have a nice shape cut into their hair or wear a natural style. If straightening is something you want to try, be sure to put your head in the hands of experts using professional products. Don't have a go at home. Chemical straightening incorrectly attempted will leave you with masses of breakage and horrible head sores. A far simpler, much less painful solution for frizz control is to apply a hair lotion by hand, to smooth down the rough edges.

Negroid hair

A very special hair type indeed. For a start, the hair rarely grows very thick. To the square inch there are less hairs than for most other types, though the extreme frizziness would suggest that the reverse is true. Growth seems slow because of the crinkly nature of the hair but some strands can actually be several inches long. Curved follicles give the spring-like action which makes most of the curls appear over-short.

This degree of curl is one of the chief differences between Afro and European hair. European type curls are uniform in shape while the Afro sprouts in numerous directions and falls in uneven waves. For many black people the answer to this 'crinkly' problem is chemical straightening: while we suggest that professional expertise is advisable when straightening frizz, for Negroid hair we stress that it's imperative. This hair has an internal structure of such fragility that each single strand experiences a tremendous amount of strain and tension during treatment and is profoundly weakened by the chemical activity. With each consecutive application, the scalp's natural tendency towards dryness increases, and there is a serious risk of damage to the hair's entire fabric. You'd be wiser to think again and forget this route to destruction. Fashionable alternatives abound: wear finger waves or sport an *au naturel* style; change colour, maybe going as far as introducing a couple of wildly extravagant colour flashes; go traditional, but with a twist – ethnic plaits and weaves in twentieth-century style are something else. The point is, you've got choices and straightening doesn't have to be one of them.

Growing pains

Just as a vegetable garden has good and bad patches, so does hair. Growth patterns can be unbelievably inconsistent and set up stumbling blocks for styling all the textures except the Negroid. Hair designs available are thus greatly determined by 'how the land lies', and whether or not you've escaped the little quirks Nature so loves to bestow. We've mentioned a few already but here's a couple more that may just come between you and that style you've set your heart on.

A *cowlick* is the name given to an area on the hairline where hair springs up in a sort of circular loop. It's a bit like having a crown at the front. There's absolutely nothing you can do to permanently tame a cowlick, and sometimes it seems you can barely control it. But you can refuse an exceptionally short cut which makes the hair stick straight up in the air. A longer line brings fullness over the annoying spot and camouflages the culprit. Additional length and volume counteract the irregular movement. If the cowlick is in your fringe, there's no way you'll get it straight and smooth because the hair clump is growing against the grain, so you should pick a style that incorporates a front bump, like a smashing curly look. Cowlicks in neck hair jump away from the neck and become spiky if cut too short.

A *widow's peak* is the very fine point of hair that grows directly in the middle of the forehead, invariably backwards, making forward styling incredibly difficult. Hair is best worn in a soft movement to one side, with a centre parting. Other styles are feasible if you leave hair exceptionally long in the front so that it can be permed to give volume and kill the push of the peak.

A *weak hairline*, where there's no strong positive growth, needs a fringe for softness around the face. Partial perming helps the hair look thicker.

Know Your Type!

Hair may also grow more abundantly down one side than the other, or fall low on to the neck in weak clumps, or grow from behind the ears forward on to the face, and so on. Such trouble spots can't be cured but they can be integrated into a suitable style and that's what to aim for.

Hair texture is not one of life's constants. Ageing and body changes give rise to alterations. Youngsters and teenagers generally have hair of excellent quality and it's at puberty that the first signs of change become evident. These changes continue until past middle age; coarse hair may be finer then and thinning a reality for both sexes. Pigmentation may change as the skin's does. In fact, by the age of fifty your hair may bear no resemblance to its former self. Diet, health, sleeping habits, and for women, menstruation, pregnancy and the menopause, plus general lifestyle, all leave their mark. The more you are in control of your own hair, the easier it is to cope with any changes in texture which might occur.

3 Going for a Clean Image

Nowadays, the humble shampoo has taken on the role of cosmetic agent as well as cleanser to remove grease, dust and loose cells. People's expectations of shampoo are now so high that if good results are not immediately evident after use, they change product sooner than question their own way of using it. Media advertisers try to persuade us that only by using their brand of shampoo will we be able to obtain the quality of hair the girls in their ads have. Well, a shampoo can't alter hair quality but with some basic groundwork, including finding the right product for your type, you can get a step closer to having the hair you want.

All about shampoos

Soapless or synthetic detergent shampoos are those most commonly used today. They lather easily, are non-irritant, rinse well from the hair and don't normally damage the hair structure. Most comprise 65 per cent water and some or all of the following: chemicals; preservatives; dye for the colour;

thickening, wetting and softening agents; foaming properties (all decent brands lather whatever the water type and whether it's hot or cold); conditioner; fragrance; and herbal or protein extract. The major differences between them include the amount of fragrance content, their powers of cleansing and conditioning, and the role they're designed to play, for example, to fight dryness, or help chemically treated hair, or clear dandruff. Several types deserve specific mention.

pH balanced shampoo

This is mildly acidic and therefore compatible with the hair's own chemical structure which is also slightly acidic in make-up. The letters 'pH' refer to the acid or alkaline level of a substance which is estimated on a numerical scale between 1 and 14. Below 7 denotes acidity and over 7, alkalinity. Shampoos range between 5 and 7 except for medicated shampoo which is about 7.3, neither acid nor alkaline. The pH of hair itself is around 4–5. While the pH element of a shampoo is important, you must recognise that it can only perform the job it was designed to do if all the other ingredients in the product are also compatible.

Conditioning shampoo

Only you can decide whether a shampoo incorporating conditioning properties can do the work of a separate shampoo and conditioner; the point to remember is that a conditioner has a totally different function from a shampoo.

Protein shampoo

The potential of this shampoo type is highly overrated. Protein cannot be absorbed by the hair – it has to be eaten in the diet. Basically, this product coats each strand of hair with a protective layer and bonds damaged hair, which then appears thicker.

Powder shampoo

This is a bit old-fashioned, messy, clogs the scalp and is best kept for dire emergencies.

What about price – how much does that matter? Like most things in life, you get what you pay for. If you pay for washing-up liquid, that's what you get. But to keep costs down generally, use sachets or trial sizes until you know you're on to a winner.

Shampoo dos and don'ts

Dos

- Judge by the results and if a shampoo cleans hair and leaves it shining bright, you can call this brand good for you.
- Look for a product that rinses away easily and doesn't leave dullness or stickiness behind.
- Use a mild shampoo that will leave the natural oils just where they should be.
- Because the number of shampoos on sale is vast, making a choice is often down to trial and error. Go through the different ranges until one seems right and if you're totally confused by the various categories, consult your hairdresser who may have a retail size available of the line she uses on your salon visits.

Don'ts

- Never use pure detergent, washing-up liquid or soap as shampoo substitutes; they're strongly alkaline, and thus conflict with the hair's own pH.
- Avoid family favourites. There's more than one hair type in a household.
- Avoid bottles with large openings because half the shampoo goes down the drain.

- Don't keep to one variety. Over time a shampoo becomes less effective because the scalp has built up a resistance to its ingredients. We don't know why but maybe familiarity does breed contempt. Alternate shampoos to allow also for the hair's quality changing from time to time.
- Don't throw out a shampoo that's a poor latherer; perhaps it's meant to be, or possibly the hair is so dirty that surface tension is too great and reduces the shampoo's effectiveness.
- Don't make your choice on the basis of pretty packaging, colour or perfume, or think 'thickness' is proof of quality – it could just include a great deal of salt.

How to wash hair

To look good hair needs shine. For shine hair must be clean. To be clean hair must be free of grime and excess oil, and this is where regular shampooing becomes vital, especially for city dwellers and commuters. So what could be easier than a good wash? A lot it seems. What appears to be relatively straightforward is really quite involved when done *properly*. All too often people confuse lots of suds with cleanliness and a job well done. But there's quite a lot more to shampooing than just hanging your head over a basin, pouring on water and dollops of shampoo, and creating an 'instant whip'. Let's face it, if that was all there is to it, there would be fewer dull heads being complained about. No, it's all a question of 'technique'. Here's the schedule to follow after you've gathered together your brush, shampoo, conditioner, a comb and a towel.

1. *Brush with a good quality brush*

Work through hair thoroughly but gently to remove all traces of dirt, dead skin and product deposits left on the scalp. Remove tangles, working from the ends to the roots. Untangle hair by holding small sections as you work up the strands, from the nape to the front, to prevent strain being put on the roots. Don't even *think* about wetting hair until all knots are out or you'll finish up with a matted mess.

2. Where to wash

The best place is under a shower or hairspray attachment. This is the easiest way to control the water temperature and ensure the flow is consistent for adequate rinsing. Adjust the water flow before you start washing because a certain water force is necessary for proper rinsing. If you prefer using the jug and wash-basin method go ahead, but try to kneel on a chair over the latter. Arching one's back for any length of time can be uncomfortable and tiring, and you may be tempted to do a rush job. What about shampooing at bath-time? We don't particularly recommend this way but if it's the best time and place for you, do it before soap and bath oil go into the water.

3. *The best way to wash*

Place a towel on your shoulders to catch stray drips then wet hair all over, taking the water through the strands with your fingertips to give them a good soaking. Hold the spray close to your head so that water penetrates to the scalp. Shampoo slips over the hair more easily and lathers faster this way. Never pour on hot water because you can burn your scalp and activate the oil glands further, especially if hair is greasy. While the water temperature doesn't affect the end result, it should preferably be warm. In a hot climate, a cold final rinse to close up the pores and refresh the skin can be fun.

4. *Putting on shampoo*

Read the manufacturer's instructions carefully. Some shampoos need to be left on for a certain time-span. If you have excessively greasy hair you can shampoo 'dry' first, to mop up some of the oil, and then wet the hair for the second application. Put a small drop of shampoo in the palm of one hand, rub your hands together and apply liberally through the hair – first with the palms followed by the fingertips. Begin at the top of the head and work down, lightly massaging to produce lather and to disperse the shampoo over the scalp to clean that too: if you live in a soft water area use less shampoo because this kind of water allows the product to lather more easily than does hard water. As you work, concentrate on the hair strands because that's where the dirt is. Don't rub, drag or pull hair and avoid scratching the scalp with your nails. Pay special attention to the hairline area – at the neck and at the front around the face where perspiration gathers and cosmetics get trapped. Treat the ends with care – they are very fragile. Don't aim for too much lather as hair can get 'lost' in the bubbles and be hard to control and wash well. Some clients ask us if two shampoo applications are necessary. Hairdressing opinion differs on this one. We'd say yes – one to get rid of modern day grime like smoke and petrol fumes, and a second as a 'finishing' cleanser. However, if you shampoo every other day as many do, one very thorough lather is sufficient. If you wash daily then only put on a light application each time as too much shampoo can cause roughness and dryness. Take time when washing your hair.

5. *Rinse it all out*

Poor shampooing results usually stem from poor rinsing. If you use a jug and basin (a bit old-fashioned and unnecessary when the far more efficient shower spray is available), you will probably have to change the water three or four times before hair is properly rinsed. Make sure water flows from the nape to the front hairline so that scum and shampoo go into the basin. The water is clear once all soapiness has gone. Some people use the 'squeak' test to check for cleanliness but it is not 100 per cent reliable because some shampoos can leave hair 'squeaky' clean without taking the bubble out of the squeak! If using a shower head, hold it close to the scalp to force jets of water through to the underneath sections, and stroke the water over your head.

6. Conditioning

We would say a conditioner is necessary after every shampoo because we don't believe in waiting until hair is out of condition before conditioning. Some clients say they don't always have time every wash to wait for the conditioner to do its job. Frankly, for the few minutes involved, this is a lame excuse. Conditioning after every wash gives hair its essential protective coating. Conditioner is combed gently through hair with a wide-toothed comb after towel drying. Hold the ends and work upwards. Leave on as long as makers recommend and then rinse. Towel dry a second time.

Hair washing questions and answers

Does combing wet hair cause breakage?

Hair breaks for two reasons. Each head of hair has a natural break-off point. If you try to grow yours longer than it is naturally able to, breaks result. Abuse by you causes breakage: using cheap tools; having hair bleached, permed or coloured too often; not conditioning; using the wrong sort of combs and brushes on processed hair when wet. Hair is more elastic after chemical treatments and more pliable when wet – by pulling you over-stretch the strands which makes them snap when they dry. Healthy hair won't break when combed wet if you work from the ends with care. Never work from the front to the back because you will increase tangles and defeat the purpose of the conditioning that has gone before.

Hard water is bad for hair. True or false?

Hard water is said to be hard work because a lather is difficult to obtain. However, we have both worked in dozens of different situations with hair and water, at home and abroad, and we have never found that the water type has made the slightest difference to the fairly balanced mix of good and poor quality hair found amongst the races. This suggests water does not unduly affect condition.

How often should you wash hair?

There are no hard and fast rules. Your hair will tell you when the time is right according to hair type, length, style, where you live and work, and the season.

Greasy hair worsens if hair is washed too often but it should not be left to get tacky as dirt particles might adhere to the grease, work under the cuticle scales and cause breakage. Excess grease encourages spots and dandruff, and makes you feel unattractive, but if possible, try not to shampoo every other day. Wait three days and use a mildly acidic shampoo, keeping your fingers away from the scalp and cleaning mainly the hair. Condition with an oil-free brand and only on the ends. Make the final rinse a mildly astringent solution, adding lemon or vinegar to restore the hair's normal acid balance. If, for some reason, greasy hair suddenly becomes dry, treat as for dry hair. If ends are dry but roots greasy, treat the two problems separately applying a more astringent shampoo to the roots, and one with lanolin to the ends. Where bad dandruff is present too, use the harsher anti-dandruff shampoo at intervals with regular shampoo.

Dry hair is low in moisture and on a scalp usually low in natural oils. It should be washed with a non-drying shampoo (cream based and containing oil to replace the missing natural elements). Wash about once or twice a week, eat oily foods and massage the scalp to stimulate the sebum secretion. If dryness stems from perming or colouring, use a shampoo formulated for processed hair. Take extra care with the ends, which are often very dry. Always condition, and in addition put on a deep conditioning pack from time to time.

Normal hair receives a balanced flow of lubricating sebum and can be washed daily if you choose but don't over-wash as you may create dryness and a flyaway problem. Use mild shampoo and conditioner.

Going for a Clean Image

Hair length also dictates washing frequency. Long hair stays clean longest because sebum works down the shaft less quickly than on short hair. Central heating, wearing winter hats, commuting daily or working in a dusty atmosphere can make hair need washing more frequently.

4 *Be a Shining Example*

Hair in good condition is like someone in good health – instantly recognisable. Just as a very fit person has more zest for life than most of the rest of us, so has hair in peak condition got that extra bounce and vitality that makes it stand out in a crowd, and easy to manage. But like everything worthwhile, it has to be worked for.

To have good-looking hair means, among other things, conditioning after nearly every shampoo. This is the only way to guarantee that the cuticle scales are smoothed down after shampooing, to allow light to reflect off their surface and to produce the shine associated with clean, healthy hair. As conditioning does not last from one wash to the next, hair needs regular treatment or it becomes dry, dull and flyaway. This is because the scales which are raised during shampooing stay that way if conditioner is not applied and allow the hair's natural moisture content to escape. Consequently, no hairdressing technique can ever achieve a first class result on such a poor base.

Conditioners on the market

The very first thing to make absolutely clear about all conditioners is that none of them can bring dead hair back to life!

Hair Matters

They are not miracle workers. What many of them *can* do is to reduce the friction between hair and the styling tool being used, by depositing a thin film of wax on the hair to create a soft, glossy, controllable material to work with. Other more intensive treatments available venture into the areas of repairing and restructuring the hair and are only employed in cases of severe damage. Below we list the various categories of conditioning agents you can choose from depending on your hair's needs.

Conditioning rinses

These have no conditioning function as such. They untangle hair and are applied straight after shampooing to coat the hair's surface with a thin layer of wax, oil or lanolin. This coating flattens the cuticle, seals in moisture and decreases surface tension so that static is reduced and hair is left sleek and shiny. A rinse after every shampoo is a good idea for most people; the modern, lighter formulations benefit greasy hair too because they do not leave a horrible, heavy feeling behind. Rinse the product off well unless it is of the 'put-on, leave-on' variety.

Conditioners

Creams, waxes, balsams and conditioning setting lotions fall into this group which have a deeper action than rinses, with some giving hair extra body and others restoring texture and suppleness to damaged hair. They are left on for a couple of minutes usually. If a product is designed to be left on but hair feels sticky afterwards, rinse until it doesn't. With greasy hair use an oil-free brand.

Restructurants

Designed to help problems of a more internal nature, they work on the cross-linked keratin chains that make up each

A classic example of healthy, one-length hair

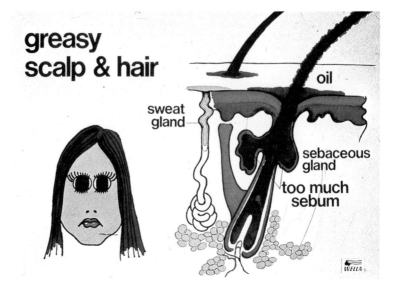

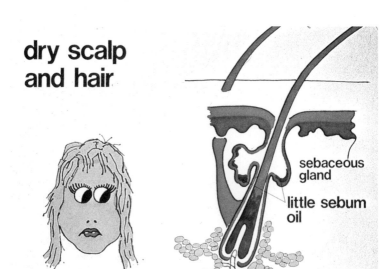

hair strand. When the links break the hair has little elasticity or strength and is weak and lifeless. A restructurant gets to the heart of the matter and attaches itself to the keratin structure where it has a revitalising effect. Sometimes the restructurant simply binds the links, a bit like a zipper in action.

Treatments

These are heavier preparations that offer much more than just basic conditioning properties and are only available as part of a salon service at the hands of a professional hairdresser.

Henna

Neutral henna is widely known for its conditioning powers but it might be best to use something else on dry hair, which henna tends to dry out further.

What you choose to use is a matter of personal choice but to begin with go for well-known brand names. If the task of choosing from the numerous brands is too bewildering, ask your salon for advice.

Application

Shampoo, rinse, towel dry. Squeeze out excess water that may dilute the conditioner. Position hair as it would lie when dry and with finger-tips or a wide-toothed comb, apply conditioner to the top and underneath sections, concentrating on the ends. Wrap hair in an absorbent towel and leave conditioner on as long as recommended. Rinse well to avoid hair looking dull and dirty, and towel dry. The big thing to watch out for is compatibility between the conditioner and the shampoo you use. Sometimes one product fights the other, so check yours get on.

Get into condition

Let us now put you in the picture regarding your own hair's condition. Your own 'condition' type is usually inherited but falls somewhere within the categories of normal, dry or greasy. What type you are depends on the moisture and sebum content of your hair. (Sebum, you'll remember, protects and lubricates the hair and scalp.)

Normal hair is normal because it is trouble-free, or almost. It has sufficient natural oil to counteract the ravages of modern living and still look good. Adequate lubrication reduces static electricity so the hair is easy to handle. Normal hair is smooth, rich in colour, strong and scurf free. But let us say right now – very few people have 'normal' hair in top condition. And if you do, lucky you, but remember, regular shampooing removes natural oils as well as grime and dirt. Use of a conditioner is just as important for you as for anyone else.

Greasy hair is caused by over-active sebaceous glands and is an internal, scalp-related problem. Too much sebum separates hair into lank, oily strands which adhere to the scalp. Greasy hair dirties quickly, especially if short, and long locks are limp and shapeless when the strands cling together. The extra oil production can encourage dandruff if the area around the glands, at the base of the hair, becomes infected and flakes. How do you know if hair is greasy? This isn't as silly a question as it may at first seem. You see, there is a hair type which is 'glassy' and shiny and gives the appearance of being covered in grease – the Scandinavian silky, flaxen hair type is one such example. Sometimes fine, thin hair which falls flat to the head can look greasy when really it lacks body. The best way to judge greasiness is to run your fingers through your hair, and if you feel oil on them you do indeed have greasy hair.

Reasons for greasiness are numerous but include changes in the body's hormonal activity at puberty, pregnancy and menstruation; stress and shock; taking drugs for medicinal purposes; bad diet and over-stimulation of the scalp.

Is there a cure? The short answer is no, you cannot get rid of a greasy condition entirely, but you can check the problem.

Be a Shining Example

1. Eat well and cut out foods that over-activate the glands such as fatty or fried dishes, spices, nuts, seasonings, chocolate, dairy products and alcohol. Eat lots of protein, fresh fruit and vegetables.
2. Try not to wear hats or scarves.
3. Opt for an uncomplicated style, preferably short, which is easy to wash. A perm or tint would counteract the grease to some extent because of the general drying effect of a chemical process.
4. Use a fairly gentle 'greasy-hair' shampoo and don't wash hair too frequently or you will create a 'Catch 22' situation where grease production is increased.
5. Never massage the scalp, and use conditioner on the ends only. An oil-free one is best but if the ends are dry use a cream one that leaves no film.
6. Towel dry rather than blow dry but if you do blow dry, use just the warm setting.
7. Keep brushes, combs and pillowcases clean. Don't style or play with your hair too much and use a comb in preference to a brush. A little hairspray will give essential 'hold' in emergencies.

Dry hair, as the name suggests, suffers from a poor supply of moisture and sebum and tends to be brittle, split and almost 'crackles' on touching. The hair has no body, is flyaway because of static, lacks elasticity and tangles easily. Both the hair and the scalp can be dry at the same time or separately. When hair is dry it may be due to inherited factors or to abuse by you. Many people think they have dry hair when they really have dry scalp. Usually if you have a general dry skin problem the scalp will reflect this also.

A combination condition can exist where the scalp is greasy and the hair dry. How do you tell? Usually the roots are greasy and flat, and a style won't hold, while the ends are dry and split. Regular brushing moves the oil down the hair shaft. Treat this hair as mildly dry and wash with a gentle anti-dandruff shampoo to clear scaliness.

While dryness can be due to sluggish oil flow, often it is a direct result of abuse inflicted on the cuticle by you. Constant use of styling tongs, dryers and heated rollers, and frequent perming, tinting or bleaching all dry hair out. Harsh shampoos strip away natural oils and allow hair moisture to escape, while other drying agents include lacquer, sun, salt water, wind, faulty diet, tools used too hot, and some drugs.

Is there a cure? There's no such thing as one solution for any problem but there are ways to alleviate it. In this instance, use a moisturising shampoo and a conditioner with oil and lubricant, which will help fill the holes created by the scales missing in the damaged cuticle. Shampoo once or twice a week at most. Avoid drying atmospheres – centrally heated or smoke-filled rooms, and saunas. Eat oily foods and foods high in protein to combat the hair's brittleness. Cut down on your use of electrical equipment – operate your dryer on the lowest setting and try not to use tongs or heated rollers. Massage the scalp to improve circulation. If chemical processes are to blame for dryness give hair a rest from them for a while and change your hairdresser, because they should have known better than to have let this happen. There are enough aftercare products available to avoid it. Of course, if you have been indulging in some do-it-yourself then you know who is to blame, and the remedy.

Have hair trimmed often to avoid split ends advancing up the shaft. Apply pre-sun treatments before sunbathing and have the occasional intensive conditioning treatment. If the dryness is exceptionally bad, you may need to make a visit to a trichologist.

5 Good Condition – More Than Just Conditioner

Now that you have your hair in good condition, the problem is to keep it that way. Using the right conditioner simply is not enough. You must eat well too, use your drying and styling tools properly, and learn how to look after your hair when it needs special care, for instance, when you go on holiday.

The importance of diet

Think about what you eat. Is your diet balanced? Do you take daily quantities of protein, vitamins and minerals, the nutrients essential for the upkeep of the body? (Refer over the page for sources of these.) If not, your body is unlikely to be healthy, so how can you expect your hair to be? Hair is made up of the protein keratin so poor protein consumption could lead to slow growth, for example, because you are short of 'building blocks'. Good blood circulation to the follicles is vital for healthy hair, so lack of a mineral like iodine in the diet could lead to hair loss. The point is this: good eating habits are as important to your hair's health as they are to your body's.

Checklist
Sources of protein, vitamins and minerals

Protein
Meat, fish, poultry, milk, eggs, cheese.

Vitamins
Vitamin A–carrots, tomatoes, oily fish like herring, mackerel, sardines.
Vitamin B-complex–Brewer's yeast, wheatgerm, whole grains, red meats, nuts, beans, eggs, milk.
Vitamin C–fresh fruit (especially citrus), vegetables.
Vitamin D–egg yolk, margarine, butter, natural low fat yoghurt.
Vitamin E–eggs, fats, vegetable oils.

Minerals
Iron–red meat (especially liver), pulses (i.e. beans, peas, lentils).
Calcium–eggs, milk, gelatine.
Iodine–fish, kelp from seaweed, iodised salt.

Hair also reflects physical changes in the body. A woman's hair can be affected during menstruation and pregnancy. Just before her period she is often very nervy and tense and some clients find their regular perm or set doesn't take so well or last as long if done at this time. On the other hand, pregnant women often find their hair takes on a new lease of life prior to the birth and it's only afterwards that the headaches begin. The female hormone, oestrogen, increases during pregnancy and interrupts the hair's usual growth cycle so that less hair falls than normal and hair is temporarily thickened. After the birth the hormone level decreases and hair that did not fall before now comes out in clumps. The entire hairline can actually be lost in this way but, happily, the normal growth pattern is usually restored before the baby is a year old. However, the new hair might be harder to control for a while

Good Condition – More Than Just Conditioner

as it takes time (along with the body), to get back into shape.

Some drugs, the contraceptive pill, illness, worry, shock and overtiredness can all have adverse reactions on hair. Just being run down, or struck by a bout of flu, can affect hair's manageability, so do tell your hairdresser if you have recently been ill because it can make all the difference to whether your new perm or tint takes well. The pressures of modern living also mean a degree of stress is inevitable in most people's lives. This tension registers on the hair and if severe enough, can caue *dandruff, psoriasis* or *hair loss*. What causes *dandruff* is still a bit of a mystery but certain factors contributing to the condition include stress, nervous tension and hormonal imbalance. Large grey or yellow oily flakes signal its presence and can be unsightly, itchy and sore. The scalp is usually greasy and unhealthy-looking. Dandruff needs to be held at bay by diligently washing hair, tools and pillowcases, and brushing shoulders frequently to get rid of any fall because you can reinfect yourself. To loosen flakes, brush well and massage the scalp before washing with an anti-dandruff shampoo (which doesn't mean a strong shampoo).

What dandruff is *not* is the shower of small white flakes you see on your clothing from time to time. This is scurf – dead skin that your scalp has shed naturally in the course of the top layer of skin being renewed. Occasionally the scaliness is unrelated to this process and is a result of careless rinsing; the scalp lacking moisture due to use of a harsh product; an allergy to a product; or the atmosphere being too dry. Scurf is not infectious and will clear up with a medicated shampoo.

Non-bacterial dandruff is a nuisance rather than a serious condition, may clear up periodically but can be around a lifetime, because it is not curable. If the scalp weeps, see a doctor or a trichologist – it could be *psoriasis*. This is a scalp disease, often inherited, related to the nervous system. The scalp becomes inflamed, tender and sometimes weeps, and can only be treated by the medical profession – ideally a dermatologist for the skin problem and a psychiatrist for the nervous one. Psoriasis is not contagious.

Between 100 and 200 hairs a day is the normal amount of *hair loss* to expect. If clumps come out or balding patches appear, hair may need medical investigation. There are instances, however, when greater hair loss is experienced as part of life's cycle. At puberty stress can produce more than the normal degree of male hormone, androgen, in women so that their hair thickens. Later, when the hormone balance is restored, the hair returns to normal, leading clients to believe they are experiencing hair thinning.

At the menopause the female hormone, oestrogen, is produced in smaller quantities than usual so new hair growth is unlikely for women in their middle years, making existing hair seem thinner.

Drugs and the contraceptive pill both produce loss but at different stages of ingestion. While loss occurs during drug consumption, (growth resumes when the drugs are withdrawn) with the Pill, hair falls after women come off it, when their body mimics the post-pregnancy period already explained.

Poor diet, anaemia and illness can cause temporary loss; over-use of perms, tints and bleaches causes breakage but the roots remain intact. Constant rough brushing, and wearing of tight plaits, roller sets and ponytails, causes baldness at the front hairline and temples. The remedy for each problem is simple – do the opposite of what you've been doing wrong. Treat all thinning hair gently. Comb instead of brushing and eliminate the use of dryers and tongs.

The severest form of loss, *alopecia*, is brought on by stress or shock and shows up as bald spots or total 'fall-out'. Regular massage and relaxation exercises, and trichological advice, should help restore the bulk lost. Never forget the trichologist is your link between the salon and your doctor, and for a contact in your area write to The Institute of Trichologists, 228 Stockwell Road, Brixton, London SW9.

How to use your tools correctly

The best way to express our feelings about the kind of hair equipment we think you should buy and when to use it, is to say 'You get what you pay for', and 'All things should be done

Good Condition – More Than Just Conditioner

in moderation'. There is no point in spending hard-earned wages on regular salon visits if at home you use a cheap, nasty comb with half the teeth missing, or you dry your hair for so long that it is almost burnt to a cinder. By following a few simple rules you can avoid wreaking such havoc.

Since their invention, *heated rollers* have become a girl's best friend. Models, actresses and other professionals who change styles often, would be lost without them. They produce a good set when and where you want, and it is comforting to know that if your boyfriend rings for an unexpected date, or your husband says he and his boss will be around in ten minutes for dinner, you can reach for your wonder rollers and perform a minor miracle in next to no time.

Unfortunately, their sheer convenience creates other problems. They become addictive and a hard habit to break. If you are already caught in the trap, remember that heated rollers were originally designed for use as a styling aid between salon visits only, and that to use them more than twice a week will dry hair terribly. Also, watch how you wind on the rollers because the spikes to secure them can break fragile hair. Position and remove them without pulling on the hair, and don't let them get too hot before putting them on.

Choose *setting rollers* carefully. Those with plastic spikes, or with hard brushes inside, tear the hair when you remove them, especially if you have used a strong setting lotion as well. Look for foam rubber ones or those with a net or suede finish, or anything else that looks as though it would be kind to hair. Avoid metal models like the plague – they overheat the hair.

Tongs, or curling rods as they are sometimes called, are helpful for styling long hair but take care how to employ them because you can quite easily burn your skin. Those with a safety trip to avoid skin contact and a thermostat for temperature control are an excellent buy. Like heated rollers, use them only to give hair a boost between shampoos; if you need them more often, why not try a perm?

Some ladies use tongs to straighten hair and this is when real damage can be done. Constant pulling of hot tongs through hair to straighten it will cause breakage. For the same reason, don't hold tongs on any one section for more than a couple of seconds and don't take them anywhere near chemically treated hair.

Hand dryers are frequently used at temperatures you couldn't bear on your skin and yet people never think of what they must be doing to their hair – creating split ends, drying it out, and so on. When you shop for a hand dryer look at models specifically for home use – they should be light, easy to manoeuvre and with adjustable heat settings. A heavy dryer is not to be recommended because it tends to tire your arm more quickly and, without your realising, can actually finish up resting on your hair. The centre back area of the head suffers most from this problem. Don't hold the dryer too close to the head and try not to lean it on the hair. Too much weight applied during blow drying will snap wet hair as it contracts when it dries. A tip for speeding up the drying process is to leave off the styling nozzle. The airflow is far greater without it. Keep the dryer constantly on the move to avoid concentrating on one area for too long.

Scissors we mention only to say, don't use them. You may be tempted to cut your own hair but we think if hairdressers are unable to do their own, then you cannot possibly do yours. A friend or neighbour isn't much better than yourself. They might cut the hair but it will look just what it is, a homespun job. Please go to a salon stylist, and if you cannot afford the full rates, ask about the cheaper student evenings when most treatments are a fraction of the normal cost and carried out under the supervision of a qualified stylist.

Brushes and combs are worthwhile investing that little bit more in, given the number of times you groom daily. For general purposes a medium size pure bristle or bristle/synthetic fibre combination brush is best. Sharp, spiky, nylon ones should be avoided at all costs as they tear hair and create split ends. Denman have a good range of easy-to-clean models with removable cushion bases and renewable ball point pins that will run comfortably through hair when you are styling. To prevent dragging and weakening hair, generally, brush from the ends up, and when blow drying don't pull at the hot hair as you unwind from the curl.

Simplicity can be the most beautiful thing

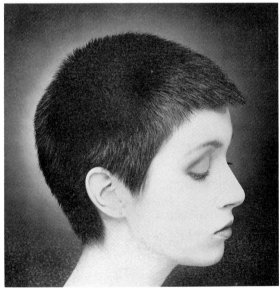

A close crop on short, shiny hair

If your comb has missing teeth, just watch how those ragged gaps catch the hair. Throw the comb away and get a new one – they are hardly expensive! Don't use metal combs or ones with a rough ridge round their edge, and look for round, wide-spaced teeth.

Both brushes and combs should be kept scrupulously clean. Too many are disgustingly dirty including some belonging to hairdressers, we are sorry to say. Shampooing and conditioning become pointless exercises if you style with a dirty brush or comb, or any other tool for that matter. Wash all equipment often in warm water, and occasionally disinfect them to get rid of bacteria, and take as your motto, 'Never a lender or borrower be'.

Backcombing damages hair unless executed correctly from the first inch of the hair root. It is really a technique for use mainly by clients with insufficient hair but who require more volume than a roller alone can give, and want a shape that lasts a little longer than a set alone would. But please note – you can leave a dreadful mess behind you if you try to tear the effect out later.

Lacquer or hairspray is another great aid to stop hair flying around but look for mild effectiveness so that you can brush it out easily, and don't apply so much that you can't do anything with your hair afterwards. Be like the girl in the TV ad: make others wonder, 'Is she or isn't she ... using hairspray?' A spray should control, not dominate, and if you can do without some during the day, do so.

Setting lotions should only be for gentle control and to give body, not strong, plastic-like types that make your curls 'jack-in-the-box' back at you and take away any natural shine you have. Choose one that will do the job without leaving an obvious coating behind.

Home hairdressing, for example, bleaching, perming or colouring at home is not something we endorse because we sincerely believe that none of these chemical processes are a job for a friend, daughter, cousin or any other amateur to tackle. Bad bleaching and perming can create the most awful, broken, unsightly hair imaginable while non-professional colouring can lead to colour build-up on your hair, overlapping and overprocessing the same hair while new hair is growing. Perms and tints alter the hair's structure irreversibly making it more vulnerable to general deterioration. To lessen the impact of chemical processing you need the professional's expertise, so for anything other than a colour rinse, go to your hairdresser.

When enough is enough Constant maltreatment leaves a mess in its footsteps – dullness, split ends, cuticle damage, and in bad cases, an inside structure that has simply caved in. Before this latter stage is reached hair can become 'desensitized' due mainly to excessive chemical processing; a restructuring shampoo and conditioner will be needed to help bond the broken layers from the inside. Often this is too little and too late, and the only 'cure' is a good cut and a promise to yourself that as new growth comes through, you will try better this time around.

Holiday hair care

The summer months can be hard on hair. Sun, sand, salty sea water and chlorinated swimming pools all combine to dry hair out. But *your* hair needn't fall victim if you prepare in advance for the fight with the elements.

At the beach

The average skin tans gently in sunshine thanks to melanin, the pigment that gives natural protection from the sun's dangerous ultraviolet rays. You help too by applying ample coatings of suntan and sunscreen lotion to prevent burning and peeling. Hair isn't so lucky. The small amount of melanin it contains is often bleached out by those same rays, and because you cannot feel you hair 'hurting', you leave it to its own devices. To get an idea of just how much sun beats down

Sun Shield cream protects holiday hair dressed in a simple inverted plait

A tousled look achieved with mousse conditioner

on hair, next time you wear a sunhat feel the top. It will be pretty hot.

So follow the golden rule that whenever possible, you will cover up with a hat or scarf. And the lighter-coloured this is, the better, because the sun's rays bounce off light shades, making you feel cooler. If headgear is impracticable or undesirable, apply a gel or sun protector as the next best thing. Daniel's clients use his Sun Shield cream which is specially formulated to condition and protect. To use, simply dampen the hair, apply the wax and make a pretty plait or ponytail, or leave loose for a 'wet-look' effect.

Water hazards

Bathing without a cap is really bad news for hair. Salt and chlorinated water can make it dry and rough. If you can't bear to wear even a nice sporty type of cap, then comb conditioner through before a swim, and pile the hair on top of your head while in the water. Afterwards, rinse in fresh water; many of today's beaches have showers for just this purpose. Shampoo daily and *condition*.

What if you spend the entire holiday sunning yourself on the beach and never go near water, should the aftercare be the same? Yes, because the scalp, face and neck areas still perspire, and salty deposits left behind need to be washed away. On top of this, a certain amount of sand always gets into hair and on the scalp, making it feel and look tacky.

Sun reactions

Sun on a perm   Permed hair is more porous than hair in its natural, 'virgin' state so the colour lightens quicker in the sun. Hair ends are most vulnerable but in very strong sunlight even dark hair can show up gingery shades. As with all processed hair, permed looks are essentially out of condition as soon as the lotion has been applied. To prevent further deterioration, coat with a sun barrier product specifically for permed styles.

Hair Matters

Sun on colour All tints fade in the sun so if you have a colour change to go on holiday, check with your hairdresser what to expect from yours. Achieving the correct colour result is hard work and time-consuming; to ruin it because of a few too many hours spent in the sun would be a pity.

Sun on bleached hair This already very dry hair can literally be drained of moisture and break off at the ends. Cover up and keep away from salt and chlorinated water.

Don't think for a minute that hair requires any less attention on a winter sports holiday. The sun might shine weakly or not at all but the strong UV rays are still at work, and the high altitudes of the mountains create a rarefied atmosphere for hair. Then there is the woolly hat you wear to keep warm that encourages hair to be greasy, flyaway and static. To remedy this, rub a little clear snow through hair with the fingers to bring instant control.

Perhaps one of the best ways to keep holiday hair in check is to wear it short. You stay cool and it is easy to clean and dry. Long hair is less manageable but can be tamed by wearing it up or held back with a headband. For simple styling, dry hair naturally. Steer clear of tongs and dryers which only increase summer damage but if you must pack them, look for lightweight, dual voltage models with transformer plugs. Pop in plenty of slides, clasps, bands and scarves for accessorising.

6 *The Secrets of a Successful Salon Visit*

You wouldn't expect to walk into the first clothes shop in the high street and find the exact style and fit you want immediately, and yet this is exactly what countless salon customers do each day of the year. Hairdressing, in terms of client satisfaction, is little different from other service industries. You have to shop around to find what suits your pocket and your preferences.

Finding the right salon

Where you find the salon to suit your tastes depends entirely on what you're looking for from a salon visit. Some clients are perfectly happy to go along at a friend's recommendation (a risky business at the best of times because no two people want the same things), while others make a crusade of tracking down the very best in their area. Here are some pointers to help in your search.

To find out what sort of places are available in your home town, start with the Yellow Pages for addresses. Read through local newspapers to see who advertises what services and look for examples of the type of work carried out. No one finds it easy to walk into an unknown salon, 'cold' off the street, so do some detective work on each place first. For example, do you know anyone that has been there already and if yes, will your source give you a fair evaluation of the service – good and bad points? Walk past a few salons and give them the once-over. Don't be fooled into thinking 'big' necessarily equates with 'good'. It doesn't. Below-average work is as likely to be found in a branch salon of a large chain as is excellent, high quality styling in a small, independent one.

What clues do you look for to help you judge a salon's competence? Start with the outside. Is the paintwork peeling, or bright and sparkling? A shoddy exterior can signify a general slackness in salon attitudes. Are the windows clean and see-through, or dirty with condensation running down them and dust on the ledges? Attention to routine but essential details indicates a thorough, efficient organisation with high standards and a professional approach to their work.

Are there any style photographs on display inside to demonstrate the kind of work being done on a day-to-day basis? Only a very few salons exhibit work they wouldn't even dare to attempt themselves. But don't be surprised to see some other salon's name credited on many of the photographs. The normal practice is for most of the top salons to show their own work, principally because they can afford to, while most of the medium and smaller sized ones purchase photographs of work done by others. Although they are able to execute the styles themselves, they just don't have the money to fund the cost of a photographic session to prove it. Of course, what really matters is whether the staff are capable of doing the work or not. To make up your own mind about the quality and type of styling going on in a salon, watch the clients coming out of it. What does their hair look like? Messy, flattering, starchy and stiff, natural? Do they seem happy? Are they all punky types or from the county set, or is there a fair mix of personalities and backgrounds? Try to establish the average age group. Some of the very trendy salons welcome all ages but might have music on so loud that you can't hear yourself think and

The Secrets of a Successful Salon Visit

only the teenagers can take the pace. Others have mainly a mature clientele where you might feel out of place requesting the latest in-fashion.

Can you be assured of any better service if there are hairdressing exam diplomas or registration certificates on the walls, or not? In one way, not at all. In this country at the present time anyone can set up as a hairdresser. Despite continued attempts by several major hairdressing organisations to persuade the Government of the day to institute a 'licence to practise' scheme, nothing has come of them as yet. Such pieces of paper give no guarantee that a hairdresser with a diploma is any more knowledgeable than any other, and affords no legal protection either. However, it is probably fair to say that any salon boss who has taken the time and trouble, personally or using his staff, to see that training is an integral part of the work timetable, really does have your best interests at heart. Many salons also display the name stickers and certificates of major hair product manufacturers, which tells you that the staff have at least received instruction in the use of those companies' particular brand ranges.

What about prices? Nearly all salons have a price list in the window or at the reception desk and this, as we know, can vary greatly in content. Study it carefully to assess what the basic costs are and what you can expect to pay for extras. Special conditioners, coffee, and the highly skilled talents of the artistic director, can all come under this label. Know from the outset the kind of money involved. After all, a salon visit is meant to be a relaxing experience. You don't want to find yourself in the embarrassing position of not having enough money to pay the bill. Nor do you want to feel cheated if the bill's a lot higher than you expected, just because you didn't do your homework first.

What about other first impressions? The receptionist is the person who makes initial contact with you on behalf of her organisation, and how she treats you counts. Is she courteous and helpful, or is she too busy shouting to a stylist that her '10.30' has arrived to even notice you standing there? Good salons train their receptionists to handle people diplomatically and in a business-like manner, so if you are unhappy about your treatment at this stage, go elsewhere. If not, look around to see some of the work in progress. Keep your ears and eyes open. Are the staff smart-looking and efficient? Is the atmosphere friendly and welcoming? What about hygiene? Do all the working surfaces and tools look clean? Are the ash trays overflowing? Are the coffee cups stained and chipped. Are the outsides of the dryers dirty? Always keep an eye on this sort of detail even after you become a salon's regular client. And don't be shy about bringing any faults to the staff's attention. They need people periodically to make objective observations, to help them maintain standards and keep them on their toes. Working in the same environment every day, they probably don't even see the deterioration taking place, whether it's to the decor or the client gowns. Most hairdressers appreciate constructive comments but in any case, as a paying customer contributing to the business, you have a right to speak up about the service and it is better to go directly to the owner than to complain to friends or colleagues. That serves no purpose for anyone.

Make a friend of your hairdresser

Go for a first appointment midweek. Nearly all salons are closed on Mondays, and Friday and Saturday are much too busy for you to receive the personal attention you'll require on this particular occasion. Be sure to be on time. There's no excuse for not being. If everyone is just five minutes late a queue soon builds up and it's certainly no fun being at the end of it. However, if you're kept waiting, an explanation or apology is definitely in order and give a black mark when neither is forthcoming.

We find people come into the salon in one of two frames of mind. Either they tell the hairdresser how they envisage their hair being done or they ask for guidance. From a professional standpoint, telling a stylist his or her job is unhelpful because you cramp the style of an essentially artistic person. By all means make suggestions, but work as a team. Such teamwork doesn't happen overnight. It takes time to become established

The Secrets of a Successful Salon Visit

because it has to be built up on a basis of mutual trust. Your stylist trusts you to tell him or her everything that is relevant to the job in hand, and you trust them to do the best for you they can. This doesn't mean that regular client/hairdresser relationships are all plain sailing. Hairdressing involves at least two minds, if not more, working together in an exercise of self expression and no matter how well you know each other, there are bound to be off-days when communication breaks down. However, if as a client you emanate confidence in your hairdresser, success is more or less assured.

Your hairdresser should . . .

1. Assess your height, frame, face shape and hair type, and study the clothes you wear for the lifestyle you lead, before ever contemplating touching your hair. He or she needs to build a total picture, to work out how the new style will fit in.
2. Find out how much time and money you have to spend.
3. Look at your hair while it's dry to see the direction in which it grows and so decide what styles would be *un*suitable. On most occasions style suggestions are made for the betterment of your hair and not only to make a sale. Professional pride makes a stylist want to produce the best possible result, which may require additional services like colouring or conditioning treatments. If you want this too, be ready to listen to any advice proffered. Hairdressers are like other business people. They know their business's long term success depends on them selling the right goods. Common sense tells them unhappy clients won't come back so they're hardly going to be foolish enough to force products on to you that you really don't need, just to make a one-time sale. It's not worth running the risk of losing custom.

You should . . .

1. Treat your stylist with courtesy. You might be the paying customer, but that in itself doesn't make you always right. Those days are gone. Most people, including hairdressers,

work best in an atmosphere of cordiality and appreciation. A little encouragement from you can spur them on to great things on your behalf.

2. Give details of what you've learned about your own hair over the years. If an awkward cowlick at the front won't lie flat, or your hair is difficult to set, say so at the beginning. They won't thank you very much if they discover the problem for themselves halfway through styling.

3. Be honest. If you're allergic to a certain brand of tint, or you're on the Pill and there's a chance it may cause hairdressing chemical processes to have a negative reaction, *tell them*. We find clients often tell half-truths subconsciously. They talk freely about what they haven't done to their hair but frequently forget all about things they have done. Be precise.

4. Have some idea of the style you'd like. Take a magazine or newspaper photograph along to the salon if that's the best way you can explain what you have in mind. Lots of people do this, though the photograph doesn't always relate to hair type requirements. Often the two are completely incompatible so be ready to accept your own hair's limitations. Don't be upset if the stylist rejects the proposed style. He or she works with different face structures and hair types every day of the week; you only study your own with the same degree of intensity. Speak up if there's a margin for discussion but do have faith in their judgement. Keep an open mind for other possibilities. The worst thing you can do is to dictate absolutely. That way you're just as likely not to attain the image you have in your head because the original one is so strong that nothing could match it. Don't follow fashion for the sake of it; adapt new trends to the present you. And if you request a style repeat, but from a different salon, don't be surprised if the result isn't a carbon copy of the first. Stylists often have greater difficulty following an already established cutting pattern than they would in originating a style.

5. Speak out if you don't like what you see developing once work commences. It's too late afterwards, when your hair has been cut too short, etc. If any of the hairdressing terminology is confusing, ask for an explanation in layman's language. The stylist will be only too happy to oblige because he or she

would prefer that you understand exactly what you're letting yourself in for.

Once the new image is complete, remember who you created it for. You. Some clients turn round and say, 'I wonder what my husband will make of this; my mother won't like me with curls; I suppose the family will get used to it'. What matters most is how you feel.

Change your style every so often so you're not in a position where you're asked, 'Did it stay all week Mrs. A ... ?' A regular once or twice a week client should think in terms of a change once a year. Opting for the same old hairdo at the same time every week, for maybe ten or twelve years, dates you. Sometimes the hairdresser is as much to blame. Staff of salons only interested in making money are encouraged to promote a quick client turnover. This is especially noticeable at weekends when they may be working on 15-minute appointments to compensate for quieter weekdays. So a person arriving regularly at 10.30 on Saturday mornings is liable to be greeted with, 'The same as usual ... ?' because it's easy and convenient. This is purely maintenance hairdressing and if you want anything more, let them know.

Unfortunately, a client can go to extremes and resort to changing salons to obtain a new design. It is very possible your usual stylist thinks a change is overdue too, but is waiting for some signal from you. Bear this in mind before making a move you may regret later on. Once in a while though, a short break from your regular people probably doesn't do any harm and, realising this, many salons now encourage clients to interchange between staff. This scheme allows a refreshing approach for all concerned and works well if done amicably and without hurting anyone's feelings.

Complaints

What's the best way to let a salon know you're unhappy with their service? Drawing from our own extensive managerial experience, we recommend sending a formal letter. We say this because over the years we've found that most clients rarely feel comfortable about complaining on the spot. At that time they may be uncertain about how they regard their new look and keep quiet. Back home, comments from family or friends put their own thoughts into words and then they telephone to complain. We prefer to have an immediate reaction, particularly if staff negligence is involved. Like most salon owners, we would like to be able to rectify matters there and then. Once you've left the premises you can't expect too much *practical* help via the telephone or through the mail, though a letter at least places the complaint on record. Make an appointment to see the person in charge as soon as possible, to discuss ways of remedying the wrong. Please don't go to another salon first. For a start, they won't necessarily know what products have been used and at what strengths, so they could aggravate matters. But more importantly, it does seem only fair to give your salon a chance to make amends.

7 Suit Yourself!

Putting together the real you is very much like fitting all the pieces of a jigsaw puzzle in place or mixing 'n' matching eyes, noses, mouths and the other bits and bobs that make up a police identikit picture. In there somewhere is the missing link that will put the other elements in focus, and without which the rest are meaningless. No matter how beautifully styled hair is, if you don't look great all over, what's the point?

Let's face facts. None of us has everything in just the right place, in the size, colour or shape we want. We have to use our assets to offset our liabilities. With hair we can't change Nature's inadequacies but clever hairdressing can more than compensate for them. There are several cuts and designs that will both please and suit you but the right recipes need the correct ingredients. In this case, a combination of face shape and features; head, neck and body shapes; posture, clothes, make-up and personality; the function the style is going to serve, for example, for day or evening wear; your own looks, pretty or plain; and your lifestyle. Make a wrong decision with just one of these and the balance is lost for the rest. That's why you're ill-advised to insist on a specific style because it reflects what you'd like to be rather than what you are. Nothing brings a sinking feeling to the stomach quicker than ending up with a hairstyle you know is a dreadful mistake and at no little cost either. By the same token, never be talked into a look that your instincts tell you is terribly wrong. Contrary to popular belief, the majority of our colleagues make it their business to *listen* to your ideas and suggestions. Of course, they can't afford not to because a dissatisfied client can easily go elsewhere, but it's much more than that: hairdressers are artists, people who care about their achievements and more importantly, want to make you happy. Hairdressing is a caring, sharing business so don't be misled by the relative few who push ahead with their own styling inclinations and leave their customers in the dark until it's too late. Communicate your view of yourself and talk through the changes you envisage. Keep a runaway stylist in check, politely but firmly. Half the time clients have only themselves to blame because they give out an unclear message about what they really want.

No-one knows your face and frame as well as you. Over the years, experiments with rollers and tongs, straight and curly styles, have taught you what makes you feel good and what makes you feel bad, so don't let someone else tell you otherwise. The few hairdressers who pressurise clients into accepting their own tastes, probably feel that if they don't leave their mark, they can't be said to be 'hairdressing'. This is their problem. Try not to get caught up in it. Wiser salon staff realise competition is stiff, and that there are a lot of other salons eagerly waiting in the wings to please their clients. If feasible, they'll bend over backwards to achieve the style you've set your heart on.

Down to basics

Before a salon visit, take a long look at yourself in a full-length mirror. Study your entire frame and gauge what balances your physique and features, taking into account hair's separate qualities. Make a top-to-toe analysis. To more accurately assess the face shape, outline your mirror image with a felt-tip pen. This works best with hair taken back off the face. Armed with details about your hair's general tendencies, your height, size, build, fashion preferences and style choice, and after

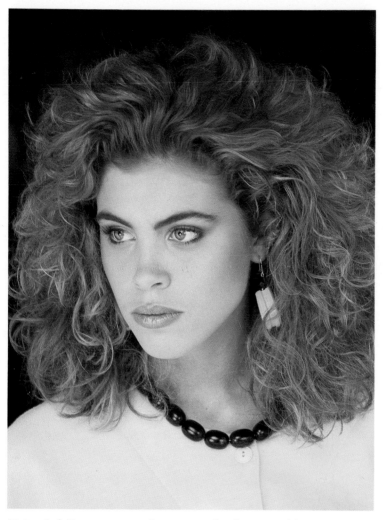

Natural, full movement enhances angular features with vegetable colour used to complement her complexion.

A square face is flattered by wispy hair styled forward with fullness.

For best effect on a pear-shape face hair is brought forward from the hairline in a textured bob.

The individual's personality becomes the styling focal point for an oval face.

A beautiful bone structure and long neck set off by an exotic ethnic style.

For a fine profile you can't beat a soft, feminine casual look.

For the rounder shape face hair is given movement around the face to create volume and texture.

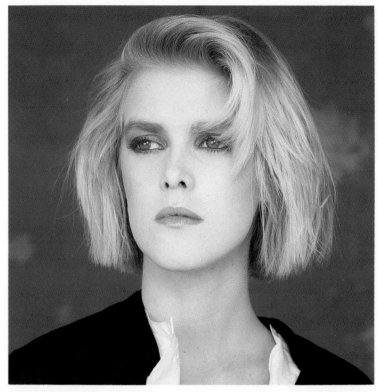

A chunky bob-shape textured to give casual width to the thinner face.

consultation with your stylist, you should reach a happy medium together.

If your nose is big you'll need fullness around the face to shrink it down to size. And a squarish, angular jawline can be softened with a full look. Basically, the greater your personal awareness of the suitability of designs, the better. When you buy a dress or piece of jewellery you can try the items on but you can't do that at a salon unless you use a wig, and while colours and perms can be changed, a cut is around for several months. To make the right choice from the start, admit your faults so they can be dealt with. Ignoring them only means they are still visible and possibly appear more prominent if surrounded by the wrong style.

With a style change, the question of 'image' comes into the story. People feel strongly about their environment and workplace, and the look that's appropriate and comfortable within these boundaries. While freedom of expression and regular fashion developments break every rule in the book, each person's life is geared to certain needs and traditional typecasting can't be forgotten. Nurses, and sometimes air hostesses, must wear hair out of the way, while a barrister with hippy clothes and long hair to match won't get many legal briefs. Spend time talking to your hairdresser about the demands your lifestyle makes upon any style you wear. As well as the type of work you do, consider the time you can devote to hair care in the mornings and, on a weekly maintenance basis, your budget. Do you require a style that's practical and functional, or formal and sophisticated? Does a regular social life demand your style should be fairly adaptable or do you want a smart cut that needs minimum attention? What's the occasion, basically? Or are you simply a follower of fashion? As fashions come and go, you follow suit. Hopefully not. If your present style looks great but doesn't feature on any trendies list, who cares.

A new style may make a new you but beware of the consequences in other areas of your life. It's easier all round to just convert to a different you in the evenings. Let your hair down by all means but allow for the transformation to be reversible. Keeping up to date is fine but it only makes sense when you feel good and look the very best you know you can.

Another thing to remember is that people usually wear clothes, cosmetics and hairstyles to impress others and to feel that they count. Most women would emphatically deny this statement but if they are honest with themselves will admit they rarely make a choice solely to suit just them. We are all trying to make an impression on someone, somewhere – a boyfriend, a husband, schoolfriends, the firm's board of directors, etc. If their opinion is very important, better to think again before opting for a style plainly out of character. The trick is to find one that will fit most of your personal requirements, most of the time.

Facing facts

So much can be done nowadays to create complete illusions of form within hair, that you can have almost any look you set your heart on. But there are various hair-related considerations to bear in mind, as well as specific body and face shapes that put a degree of limitation on styling flexibility.

Thin or fine straight hair becomes messy easily so must be cut into a short style or lightly permed to give a degree of control.

Medium textured and straight hair is highly adaptable. Go for a mid-length bob if hair is in super condition, or soft perming for gentle curls. Plaiting, tonging and crimping are other options.

Fine and curly hair needs a perm to keep curls in shape or a sharp, precise cut to prevent its habit of going haywire and becoming unruly.

Thin, wispy and lank hair can't be grown long. Patience and perseverance don't pay off. This hair hasn't the capacity to grow beyond a certain length.

Thick and strong hair is hard to control. Mid-length is easiest to cope with and setting on large rollers will keep the volume in check. But hair has to be continually dampened down to reduce frizziness.

Suit Yourself!

Styles to suit body and face shapes

While texture type and growing power determine a style's suitability, body and face shapes are the overriding controlling factors. Of course, the two work in conjunction. Your head can't be seen as a separate entity. Here are a couple of pointers to think about with regard to the different body sizes.

Tall and slim

This description covers heights between 5ft 8in and 6ft. Most women of this type can take advantage of the gamut of styling variations. You are fortunate because your height and build allow for styles with volume and length which balance your frame shape and add a touch of femininity. However, hair left to hang lank and close to the head does nothing for anyone. Hair left loose should have volume and lift so roller setting or tonging is essential. Tall women need hair bulk for their head and body to be in proportion. A really short cut will only emphasize their height. Medium to long hair is best and can be worn partially tied back or up in an elegant chignon.

Tall and well-built

A big-boned frame should be accompanied by soft, voluminous hair, preferably mid-length but not long. Large people tend to surround themselves with masses of hair, presumably to make their faces appear smaller. Sadly, the opposite effect is usually achieved because additional dimensions are joining those already there. Aim for a look that cleverly combines sufficient width to frame the face with enough softness to reduce your generous proportions. Don't wear hair off the face. That's literally thrusting it forward into prominence. Ideally, maintain hair between shoulder length and chin level; shorter than that is out of the question unless you don't mind sporting an inverted lampshade figure. Shoulder-length hair works best and can be worn sleek because a certain amount of volume will be created naturally, or swept into a chignon with lots of body and fullness.

Short and thick-set

Heights between 5ft and 5ft 4in qualify here. Short, plump women should be wary of over-crowding the face frame. Too much hair is overwhelming, though volume *is* important, because the body is thrown off-balance and the overall figure shape made shorter. By far the best solution is a style with a fair amount of volume and a similar degree of lift. Chignons are fine because they create a longer neckline, back and front, leaving the neck free. When the neck is covered round the back and sides, the whole body shape is reduced and widens out. For example, left to fall in a bell shape, hair emphasizes the plumper contour. Too short hair is harsh, and a strong, dramatic cut can seem hard and unattractive.

Petite

Provided she has the right sort of hair in the first place, a petite person has a lot going for her stylewise. For her, very short, feminine soft cuts, avant-garde looks, and fluffy, controlled perms are *in*, and long or big shapes with bundles of hair are *out*. Much of what you should steer clear of makes commonsense. Hair crowding a small face will look top heavy.

Narrow top/wide hips

A certain mass of hair is required to counteract the slimmer top. Short hair will accentuate the extra width or heaviness in the lower limbs just as hair worn too close to the head will. Look for a longer line, sleek or fluffy, that has length between the chin and shoulder area.

Suit Yourself!

Wide shoulders/slim hips/narrow bust

You may wonder what any of these features have to do with hair but in actuality, overall balance counts for more than you probably imagine. If you are larger on one part of the body than on another, hair must obviously help to compensate.

Every body shape comes in small, medium and large models with varying imperfections. How seriously each imperfection is viewed depends on individual hang-ups but the right look fitted to face, neck and body basics, makes the world of a difference. Rarely does a single feature prohibit improvement. Any skilful stylist will have the competence to diminish the features you prefer to camouflage and accentuate the better ones.

Pear-shaped face

Characterised by a wide forehead, narrow chin area and small cheekbones, the pear shape needs hair to fill out the area between shoulders and chin. Avoid cuts taken short into the neck but if done, they should be brought forward from the hairline behind the ear and fixed into the narrowest point of the pear shape. Don't have hair down past your shoulders as it will just hang over them and emphasise the narrowness of your chin. And don't take hair back in a ponytail, etc., as this accentuates width at the top of the face, and thus, of necessity, the narrower bottom. Hair styles with forward movement, for example, a smooth bob or a graduated cut shaped down and out, help greatly, and a sleek look with volume through the hair is a definite asset as long as it's maintained in the lower sections too. Partial perming can be executed where extra volume and width are required.

Heart-shaped face

This is a very pretty face shape. Main characteristics are a fine chin, width in the cheekbones and a nicely curved head with broadish temples. Most romantic and chignon styles or full, high-hair looks promote these features beautifully. Hair worn away from the face is particularly attractive but is inclined to add width to the forehead. A wispy fringe will conceal the forehead's breadth and bring it into better proportion with the chin area. Discount full, flat pageboy fringes or Sixties-style middle partings which level out fullness and height.

Oval face

Regarded by most hairdressers as the nicest face shape of all, the oval is totally adaptable. The 'egg on its point' is wide and rounded at the top near the temples, and narrows down towards the chin. The cheekbones are usually strong and the eyes wide apart, so that natural good looks mean you can have any style you want. The individual's personality becomes the styling focal point rather than any physical feature.

Square face

Basically, in this case the width across the jawline is approximately the same as that across the temples. Unless you are very pretty and the owner of a tiny, square face, a short cut won't suit. With larger features the last thing in the world you need is a very short, severe style. It's best if hair falls on to the face in curls or waves, or is swept off to one side. Don't centre-part because you will then stress the symmetry of the face's angles. Treat similarly to rounder shapes: soften lines and take longer looks close to the head to frame the face. Direct hair towards the face, never away. Encourage a small amount of height in any style and if you fancy a fringe, remember it'll need to be longer at the sides than on the forehead to continue the 'rounding' effect.

Angular face

Prominent features are a square jawline, high cheekbones and a generally very attractive shape conducive to all manner of styling. While a number of people may disagree with this diagnosis, the fact remains that even when hair is worn off the face, revealing these traits to their fullest extent, they look terrific with good make-up and hairdressing. However, an exceptionally short cut makes angular lines appear rather harsh. Shorter lengths must contain softness and texture, especially if hair is contoured around the ears. A strong chignon pulled back into a reasonably tight form that leaves the face in full view, is another option. If you prefer something that diminishes the width of the jawline, hair may be taken down the sides in a sleek or voluminous design to cover it. Framing the high cheekbones will reduce the face accordingly, and sleek lines are good because they swing to and fro, hiding the shape at intervals. Wear a longer line back over the shoulders for a full, soft, feminine touch with volume at the back of the head.

Long, thin face

A thin face needs softness and width to lessen the length, and because it's gaunt and hard looking, requires the softness to fall forward to fill out around the face and neck and add height. Left straight, hair hangs the length of the face, calling attention to the thinness, and more so if middle-parted. Buns, rolls and upswept styles of any description are out. They're unflattering because, regardless of the volume built in, the thin effect remains. Perming or setting in gentle waves that move on to the sides is the answer. Short cuts are fine if accompanied by width. It's essential that the width is maintained and a width-making style, if achieved by a set, mustn't sag during an evening out. Hair cut tightly into the neck creates a shape where there's more volume in the upper part of the face than in the lower, unless the sides are made longer too. Reduce forehead depth by a soft fringe.

Plump/rounded face

A tendency to surround this shape with more rather than less hair, to make the face appear smaller, seems to work. For example, traditional Japanese headdresses involve their owners putting the hair up and encircling their face with enormous, extravagant ornamentations which appear to shrink the face and increase its length. In modern day terms, soft perming is great here if hair is straight. By encouraging waves and movement around the face you create volume, thus breaking up the roundness or chubbiness. Don't wear hair totally sleek and hanging or smooth-topped with waves on sides only. If you wear a bob ensure the fringe isn't over-wide and that hair falls across the corners of your eyes and covers your cheeks. This has a narrowing effect as well as giving shape on top. Forget about a chignon unless you are prepared to sport a fairly large one and be slightly out of fashion. A short look should be wide-topped.

Tackling problem areas

Low forehead

Many people feel they cannot have a fringe with a low forehead but that's not true. With a fringe cut in correct proportion it's impossible to tell the size of the forehead underneath, though hair should be brought forward in some way for the balance to be right.

High forehead

Options here depend upon facial features, but on a general level covering the forehead entirely is usually a mistake. A soft fringe or light tendrils work best. If you have breadth as well as height, a full fringe rather than a smooth one will serve best. On rare occasions, someone with an especially beautiful face can wear the hair taken back.

Suit Yourself!

Long neck

Maximum attention should be paid to shortening the length. Never centre-part the hair and so create massive, droopy, 'spaniel-like ears'. A softly waved or curly mid-length look will produce the necessary reducing effect.

Short neck

At all costs, don't swamp this outline with a mass of hair. Keep it short and shapely. Hair contoured into the nape will give the illusion of additional length.

Prominent temples/forehead

Jutting out temples need covering. While the best solution is to have a slightly off-centre parting so that hair camouflages the bumps, you can't afford to have a side-parting which would bring the temples into greater prominence. Where the forehead stands out, choose a sleek fringe to mask it. Conversely, to counteract a receding forehead hair must move in a forward direction.

Ears sticking out

Cover them up unless they are small and dainty.

Double chin

Rolls of chin flab can't be concealed but can be camouflaged. Avoid short shapes and go for medium lengths below ear level.

Cowlick

Whether it's on the front hairline or at the neck, if a cowlick is cut too short it won't lie flat. Some length must be left in the hair over this spot to avoid unattractive 'spikes' forming.

Weak hairline

This feature is found mainly in heads of fine hair but the lack of a strong, positive contour line can exist in any hairline. Weakness occurs most often when hair is broken off and new hair is growing through or because of irregular growth. A fringe, or softness around the face plus part-perming to thicken hair, solves the problem.

Large nose

Hair needs softness to reduce the impact of the nose's size. Never take hair off the face and so draw attention to the very last thing you want noticed, your nose.

Small face

A tiny face with masses of hair looks swamped nine times out of ten. The face's character and the profile's versatility are both lost because they're hidden. Short, neat looks; sleek, longer styles or controlled chignons are alternative ways of enhancing and enlarging miniature features.

Glasses

With the price of spectacles as high as they are today, you can't afford to ignore their effect on your overall image when teamed with a particular hairstyle. Common sense will tell you large frames don't match a neat, short cut on a small head and that small on small is what to aim for. If spectacles are something you wear every minute of the day, hair naturally plays second fiddle and styling decisions revolve around the glasses' frame shape Talk to both optician and hair stylist before purchasing a pair.

In any styling decision, the key word is 'balance'. Each and every face shape can be cleverly brought into its own with skilful cutting, setting, perming or colouring, and the coordination of make-up and clothes. The trick is to consult every expert you can lay hands on. Too many clients act first and seek advice later. If you intelligently convey information on the sort of look you want, any decent hairdresser will interpret your wishes with sensitivity and make the necessary adjustments to allow for your features. They'll be sympathetic towards your hang-ups. Sometimes our obsession with what we consider to be bad points actually turns them into 'blind spots'. We can't see round or through them and in every move we make the 'faults' predominate in our thinking. Others can help put problems into perspective. A stylist can sift fact from fiction and tell whether a style will suit from every angle, sitting or standing: the two are very different.

The whole process is a bit like doing magic. Hairdressers and make-up artists are magicians and illusionists in the way they shape and shade a new persona using angles, light and colour. That's why the two areas of expertise must be tackled in sequence. Treating them as separate entities has much the same effect as buying a blouse of a certain precise colour on one occasion and then trying to find the perfect matching skirt another time. You rarely do and are left dissatisfied with a lopsided mix 'n' match. Look upon yourself as a 'whole' before making decisions.

The importance of a good cut

If you never cross a salon threshold for anything else, go for a good cut at least every two months. The basis of any style is the cut underneath, and its excellence or otherwise serves as the framework to whatever follows. Scissors and razors are both used for cutting, with the former in widest use since the 1960s. A razor produces a cut with very fine ends and a precise contour but the degree of control the stylist can wield over hair is far less than with scissors. For complete cutting precision and perfection, the control possible with scissors is hard to match and explains why the scissor cut has scored so much in the last twenty years.

Razor cutting is always executed on wet hair to avoid the hair pulling, while scissoring can be carried out on either wet or dry hair. Scissor work is as varied as the result desired but razoring requires a real expert at work for hair to be better than merely passable. The technique is also limiting. Blunt ends are softly tapered and the ends of fine hair can be left too fine. However, while the razor cut is more enduring and has greater style movement, razors will never totally replace scissors. Both tools are interchangeable and serve the stylist's purpose on different occasions.

The art of cutting today bears no relation to that of the past. The skill and technique involved in a modern cut goes far beyond anything done before. Consequently, it costs that bit more but looks better for longer. Short hair is so convenient to wear that, by and large, it wins hands down in the popularity stakes. Short is 'sharp', as the Americans say (meaning, with oomph). To top this, 'shorts' are wearable, controllable, washable and stylable. But even for just a trim, a salon visit is vital. Don't be persuaded, however, that hair growth is encouraged by cutting. Your hair grows at a certain rate and that's all. Trimming tidies the ends and the style's overall appearance; cutting creates the initial shape and, at intervals, promotes a new you. A change is necessary now and again and we take it for granted that you won't attempt cutting your own hair or let anyone other than a hairdresser do it for you. Cutting is not merely the meeting of two blades – the angles worked must fall back into place as you go about your business afterwards.

Cutting long hair short is probably one of the most dramatic moments in salon life. A sympathetic stylist appreciates the tender feelings present at this moment of truth and is likely to suggest that the length is shortened in stages. You may want your hair kept at one length if you eventually hope to grow it long again but, depending upon the original length, this could take years to achieve at the average growth rate of half an inch per month.

Cutting terms

Blunt cutting takes place when hair is cut straight across and little curl remains on the ends. All hair can be blunt cut and blunt cuts can be graduated and layered too. Around the face a blunt cut can be complemented with point tapering, where the scissor points are prodded into the hair to thin sections around the face wherever less weight is required.

Tapering is performed on dry hair and the scissors are used slightly open and run along the hair shaft, tapering off. This method was popular in the 1920s and 30s and has the advantage of making the ends curl. In our opinion, the tapered cut lasts for a shorter period of time than the blunt one, which has a chunkier and bulkier finish because the ends are cut straight and the shape alters less noticeably. Tapered hair can be graduated and layered but the sections show up quite distinctly.

Graduating is a process of shaping hair gradually from one area to another. The first example of the technique in use was in the 1920s and was called 'shingling'. Dry hair was graduated at the back of the head, scissor-over-comb, in an upwards direction and fitted into the nape. The 'Eton Crop' was cut on similar lines. Graduation produces texture continuation and appears in looks as diverse as punk, shaggy, head-hugging and flowing. But graduated hairstyles last as much as two months less than all-one-length ones do.

Whatever way hair is shaped, it must be well done to maintain a good form. You rely on a cut for the success of future styling. New look cutting and razoring mean you can just shampoo hair and leave it to dry and still retain an excellent shape. But *the* basic requirement for all quality styling is still *healthy* hair.

Styling tips

Straight hair

- If hair is also fine avoid graduation, which emphasises the natural lack of body weight. One-length blunt cutting best provides an impression of volume. Perming or colouring has a thickening effect.
- Keep split ends at bay – they are very noticeable on straight textures.
- Avoid rough handling of hair when washing and drying. Don't backcomb fine hair. Better to create artificial fullness by drying *against* the growth pattern.
- Fed up with straight lines? Plait hair into small twists or set with rollers.

Curly hair

- You can't take the spring out of curls but you can reduce it considerably with a perm.
- Avoid knotting by drying carefully and brushing through with a wide-toothed comb or brush.
- Watch that conditioner, hairspray or gel don't weigh heavily on hair.
- Dry naturally or with a heat lamp, and on summer days add shape with gel. A gelled wet look remains until washed out.

8 Get Set

Although setting diminished in popularity with the advent of blow drying, natural drying and wash 'n' wear perms, the technique enjoys modern day revivals especially when sophisticated, classical styles become fashionable. Most hairdressers offer this service unless they are purely 'blow dry' specialists, whose interests are normally strictly limited to cut and blow work. Good setting skills take years of practice to acquire and even then not every hairdresser feels empathy for the intricacy of the finger work demanded. Before signing up for a session, you might profit from making enquiries among friends to find out if they know of anyone adept at setting. Seek out a true master of the art.

Setting techniques

To explain the concept in a broad sense, setting is putting in rollers, finger waves, tiny plaits, pin-curls, and even a perm, because all these techniques involve the placing of wet hair in a 'set' curly or wavy position. A narrower, more precise description is that it's a temporary way of shaping hair by adequately drying wet, elastic sections laid out in a specific pattern; the moulded shape remains until the hair is washed again.

Setting is basically another form of hair design but its suitability for all age groups is not always appreciated. There's long been a mistaken belief that the technique is for middle-aged customers, full stop; the prerogative of the 'shampoo and set' brigade. Not so; now, more than ever before, setting is universally an integral part of the commercial hairdressing world. Traditional clients have been joined under the dryers by younger women never previously exposed to a strong fashion trend in setting, but fast learning the benefits of a set. Advancements in styling techniques and products mean modern setting is exceptionally natural looking and not the same monotonous style put in week after week as in the past. The advantages are two-fold – increased flexibility and versatility. Changing styles simply calls for hair to be set in different directions. In fact, half the fun lies in the temporariness of it all. There's no end to the styling variations you can create for a multitude of occasions – soft, smooth waves for the theatre, cute curls for a party, or a loose roll for a walk in the country. And if a perm is something you've long dreamed of trying but hesitated to take the plunge, a set's the answer. You can have the look with none of the permanency.

How long a set lasts is dictated by the hair's texture and the support framework below it. On a perm, for example, a set could easily last two to three days and straight hair with a body perm might hold one for as long as a week. If backcombed and lacquered (but hopefully neither), the mildest of modern sprays will prolong its life still further. Use a good quality setting lotion for this extra hold, not one of those awful gummy, sticky brands. Between salon visits, pop a roller or pin clip into the spots beginning to flop to help maintain the basic style shape. Watch your stylist attentively when he or she arranges the set, for an appreciation of roller positioning. This will simplify things when you come to restyle later, though sections around the back of the head, where a certain amount of movement is essential, will always be harder to handle.

Hair is set on rags, unwound and run through with the fingers for this beautiful effect

Thick hair set loosely on large rollers, hood-dried and brushed through

In *roller setting* (rollers are best for a curly finish and pin-curls for waves), as few as five rollers or as many as forty may be wound in varying directions for different results. The size of roller differs enormously, with the largest giving big, loose curls and a soft effect, as you might expect, and smaller ones, tighter, more defined curls which look specially nice on short hair. Tight curls can be relaxed by setting them on large rollers. Roller size and angle positioning have a direct bearing on the degree of movement and shape achieved. A few rollers give sleekness while just one enormous roller, wound with the bulk of a head of hair, can produce an Evita-style roll. Normally a combination of rollers and pin-curls are used to set one head. The pin-curls are used to tackle the awkward spots such as behind the ears.

How does it all work? Shampooed, conditioned, wet hair is towel dried and smoothed down with setting lotion, when used. Lotions for all types are available in various strengths. Hair looks more natural without one but a lightweight variety certainly benefits wispy, fine, flyaway hair. The wet hair is combed into the shape aimed at and then strands parted off and eased around each roller in turn. Work is usually carried out from the front in sections, each a similar size to the roller type being applied. The angle at which each strand is parted off depends on the direction in which the hair is to flow. For example, where the sides are to move backwards, the vertical partings taken are rolled away from the face, or if the sides are to cover the ears, horizontal partings are taken and rolled downwards. The ends are placed evenly and smoothly over the roller to prevent them twisting or buckling and giving the style a slightly fuzzy, unfinished appearance. Once the ends are properly positioned the roller is wound down to the head, with any loose wisps tucked in, or if the strands are too short, pin-curled instead.

Rollers are never plonked one on top of the other and hair must have sufficient tension on it for them to stay in place. Crown hair is rolled towards the nape and vice versa. Once dry, you can clearly see how the hair has taken on the roller mould. Stiff sections are loosened by gently straightening out strands with a brush. Be sure the stylist puts in adequate rollers for the job in hand as some do cut corners. This

Get Set

frequently happened in the 1950s – a few rollers on the top, hair backcombed and the client sent away all lacquered up. By putting in less work, setting time is reduced and a higher turnover of clients achieved. So if you see loads of gaps between rollers, and know full well your last set hardly lasted a day, ask for more to be added. You're entitled to value for money and despite the temporary nature of a set, you've the right to expect some durability.

Keep an eye out for dirty or worn rollers with springs poking through. If the spiky sort that tear hair and dig into the scalp are employed, speak up. Never sit and suffer. Most stylists will be only too happy to remove the offending pins or pad uncomfortable rollers with cotton wool.

Natural drying is such a long, arduous process with a set that it isn't really a viable proposition although the bigger, rollered styles dry fairly well as air circulates around them. You'll be popped under a hood dryer for 30 to 40 minutes and your stylist will test an individual roller when the time is almost up, to check if the hair is dry. If not, the roller will be rewound and left a while longer. Before washing again after a set, brush out any remaining lotion or spray used as a styling aid. If you do this on a daily basis while the set holds, you'll prevent product residue or dust adhering to the scalp.

Heated rollers produce an instant set anywhere, anytime hair needs a gentle boost or total restyle, but the curl is weaker and shorter-lived. For best results, the hair has to be slightly damp and for this reason, plus their less drying quality, steam models with a mist spray are ideal. Leave rollers in until cool to stop the still warm, pliable hair being reshaped accidentally. Brushing or combing curls soon after forming them is liable to pull the effect out. Break up the 'set' lines by brushing hair in the opposite direction from which it's been set and then as normal.

Finger-waving is literally waves pressed into thoroughly wet hair with the fingers. A comb may also be used to form the peaks and troughs of the waves by holding hair sections between two fingers and sweeping the comb's tail up the centre. The procedure almost resembles fluting the edge of a pie pastry for decoration. For a smooth effect hair must be wet. Equal portions are groomed along the path the waves are to

Mid-length, textured wispy look moussed and naturally-dried

Short back and sides look casually worn using setting mousse

take. Deep wave lines give a strong definite shape but are only workable on thick hair. Where the base is shallow, for instance behind the ears and on the crown, waves fall out quickly. It's better when hair's own gradual progression from thinner to deeper areas is utilised than when uniform waves are sought all over the head. Natural movements are obtained when hair is combed in a gentle upward motion and left exactly where it falls.

Pin-curls complement short hair well and are so simple to make, they're almost too good to be true. Hair is worked in extra fine strands, each wound around the finger, pointing in the direction the curl will follow. By encircling it on itself, the curl can settle comfortably on the scalp and be secured with grips or pins – not clips, which leave demarcation lines and flatten hair. The curl can also be combed into shape with the ends twirled into the middle. A tiny spot of setting lotion coating each curl holds the design line without detracting from the light feeling now present.

Finally, there is a fun type of setting tool made of foam rubber that major salons sell under their own brand names, which is as easy to apply as a roller. The bendy foam pieces come in varying lengths and are wound round washed hair and then bent over to secure the hair. The result is as curly or wavy as desired.

Setting agents

Setting lotions come in different strengths – light, normal and extra hold. Usually the light to normal variants have sufficient holding power for the average head of hair and give body, shine and manageability. Most have conditioning elements, and some even help with dry or greasy hair problems or help fight off the effects of humidity in the atmosphere. To use, sprinkle over shampooed and towel dried hair, covering every inch of the head, top and underneath. Comb hair and position styling aids.

Blow dry lotions give light control, volume and shine, and are used to best advantage when applied to natural looks.

Hairspray has also kept up with progress and the layers of plasticised spray of earlier times are now no more. The lightness of the new variants leaves hair very natural looking and allows for easy removal by brushing and washing. New sprays are lacquer- and oil-free, and non-sticky. Your style will last longer too.

Gels have been in the big time for several years now but are not really a new product. This essentially jelly-like substance is a marvellous setting aid for today's trendy styles and is a much lighter and less greasy version than those popularised in the 1930s. The gels are non-gummy, non-stick, fast drying and give body and volume. They wash out easily and this is the big bonus of gels, their sheer versatility, because you can sport a spiky high fashion look at a disco one night and be back at work next day with hair returned to normal. The important thing with gel is to control the amount put on. Just a hint on the ends is enough to boost existing volume, but a heavier application is necessary for a sleek wet look which is achieved on wet hair combed into style and left to dry. Gelled hair should be left alone once styled or the holding effect is lost. For special occasions coloured and glitter gel is great fun.

Mousse 'magic' foam is the new way of adding body and bounce to hair *and* holding a set or style with gentle firmness that lasts. Conditioning agents give shine and gloss, and reduce static too. A moussed style can be revived to its fullest extent with a little moistening. Mousse is so marvellous that most salons use it automatically as their number one setting aid. Spray-on mousse can be applied to wet or dry hair. It should be shaken well before use, applied sparingly with the fingers and spread quickly before it dries. Daniel's own Blow Foam Setting Mousse with Conditioner is designed for every hair type and gives extra body and bounce. The foam is oyster coloured and is available from the Galvin salons and selected stores and chemists.

9 When the Heat Is On

Drying hair is as much an art as a practical necessity. Techniques come and go but the ultimate goal remains the same, for drying to be the means to a beautiful end. *How* you dry your hair creates that final style. Successful drying is somewhat more complicated than pointing a dryer at your head and switching on, or leaving hair to dry haphazardly by itself. It takes know-how to pick the appropriate method at any one time and skill to choose the tools for the job and to manipulate them. Learn to hold and handle drying and styling equipment just like professionals stylists do. What's the point of paying for an expensive cut or perm when you don't know how to style the shape yourself. Bubbles of air through hair give bounce and body. Your role is to artificially create them. The hairdresser lays the groundwork but between salon visits requires your assistance. You'll obtain greater personal satisfaction from their efforts if you feel in control. Know what does what, and why.

Drying techniques

Towel drying is as easy as one, two, three. Gently squeeze excess moisture from the hair and pat dry with the towel, or wrap it around the head turban style, to catch drips. Rub gently to avoid causing snags. A smart alternative is the terry towelling mitt. Simply slip a hand into the glove and wind it in and out of the hair. It's marvellously stimulating for the scalp and so effective one wonders why nobody thought of the idea sooner. Short, summery cuts feel wonderful after a brisk mitt dry and a walk in the sun.

Finger drying speaks for itself. You're continuously employing your fingers and the warmth of your hands to simultaneously mould the design and encourage drying. Take your fingers through the hair from underneath sections, and work towards the crown area lifting the hair up as you go. Finger-dry a perm to loosen controlled curls and produce full volume.

Natural drying is best left for warmer days or you'll give yourself a chill in the head. After cutting and shaping, the hair is left to dry as it lies. Easy perms and shaggy, tousled looks of recent years promoted this 'leave and let live' process. There's definitely something very appealing about shaking away watery drips and stepping out into the warm sunshine with hair flowing free. But it's not an excuse for messiness or to be over-casual. Hair needs shape and design more than ever when the finish is destined to be unstructured. And while natural drying is automatically a timesaver, a word of warning. If hair tends towards greasiness then take a wide-toothed, afro-style comb through it to reduce the likelihood of a flat and lifeless finish.

Lamp drying is a favourite technique among hairdressers but rarely used by clients at home. The beauty of the lamp is the stillness of the heat emanated. A few models also carry fans but generally the infra-red lamps generate a motionless heat which radiates out towards the hair without displacing or disturbing the style. Certain permed looks are 'set' in position, lamp-dried and brushed or finger-ruffled afterwards, to obtain a completely natural effect. But heat is heat and if held too close to the hair, the lamp will scorch the ends. Salons minimise the chance of this happening by utilising multi-lamp models, with movable arms and adjustable heads, which operate from one main stand. The sheer number of bulbs hastens along the drying process, but none of them is near enough to the hair to cause damage.

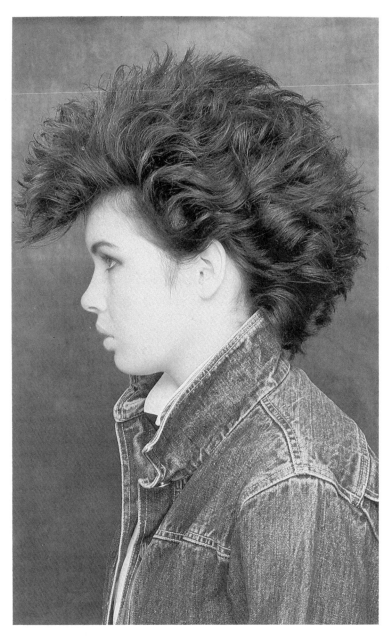

A punky look dried using mousse and fingers plus hand-dryer working against the roots

Hood drying is chiefly reserved for sets. Of the two types marketed, the 'air-swirl' hood is probably most commonly in salon use. Hot air gently wafts from its small jets and is distributed in a comfortable, even flow, eliminating the need for a net. The conventional 'blower' hood allows for less diversity of styling because the forcefulness of its air-flow dishevels hair unless a strong setting lotion and net are used. For some reason, too, client complaints about clips and pins overheating are more frequent with this machine type. Sophisticated touches in both types of dryer aim at total client comfort and include cutaway sides in the hood; an electronic thermostat providing precise temperature control; a customer-operated control switch, a face cooling unit, and built-in music and ash trays. But their biggest plus over the hand-held dryer is that they cause hair less heat damage.

For home use, the most convenient and readily available hood is the sack variety. The portable attachment permits freedom of movement during drying and, as many are battery operated, you're not even restricted to mains control. Not everyone is comfortable with a sack over their head so try out a few models before buying.

The *hand dryer* is used for whole-head styling, to direct heat wherever you plan shaping – in contrast to the hood which is principally employed for drying sections laid out in a set formation. The styling nozzle attachment concentrates air flow in one spot, so is marvellous for blow waving and to give lift, but for speedy drying of large areas it should be replaced by a diffuser head or the dryer left open-ended.

A good hand dryer is a light and manageable one, with a range of speeds and of medium wattage. For domestic use there's no real advantage in owning an overly powerful instrument. Look instead for utility extras like blow brushes and combs; or pick a handbag size dryer with foldaway handle for portability; or a dual voltage model, a must for foreign travel which is increasingly a feature of modern life.

Handling a dryer calls for abundant patience. It's tricky, tedious work and you'll be tempted to take short čuts, but don't. Test the heat temperature against your hand, retaining the highest setting for the first few minutes of drying only. Hold the dryer about a foot from your head and use it in short,

Towel-dried then gelled sides and back to accentuate front

Moussed and finger-dried forward

Cut short and naturally dried into gentle movement

With setting lotion, brush and hand-dryer a voluminous look is achieved

sharp spurts. Keep that dryer on the move. More heat damage to hair arises from careless hand drying than from just about any other source. Hair quickly reaches a drying saturation point which you can fail to spot, especially on the ends. It'll be now too that you'll strike oil as the excess heat over-stimulates the sebaceous glands. Whenever feasible, keep direct heat off scalp and roots and exert minimal pressure on the hair with your hands. Pay extra attention to chemically processed locks, which can virtually be blasted away with harsh drying.

We take for granted that your machine is dust free and electrically sound. An interior blocked by dirt particles will reduce the air supply to the fan, cause the element to overheat and may even set the dryer on fire. And don't twist the flex when storing because its casing may crack or fray and leave wires exposed. Dangerous to say the least.

Blow drying in stages

Despite the massive publicity given to blow drying in its heyday, there remains a surprising number of people unacquainted with even the basic steps. When blow drying originally made an impact in hairdressing, stylists quickly took up the challenge to create increasingly simple cutting lines for easier all-round shaping and styling. The market has literally been flooded with super-short, precisely executed commercial cuts that look smashing when people walk out of the salon, but which don't stay that way long because their owners haven't conquered blow dry co-ordination. Here's how:

To begin with, work in front of a mirror – it helps to see what you're doing! And as few of us are ambidextrous, it's logical to hold the dryer in your strongest hand for the bulk of the drying and to brush with the other. Work on towel-dried hair and, if you can, wait until the hair is three-quarters dry before commencing styling. Dripping wet hair is harder to manipulate and takes such a time to dry that heat damage is almost guaranteed. Have a waterspray handy to damp down spots that dry before you're ready to style them.

Shape the hair with a comb while putting heat-protectant lotion on all over, or just on the ends if you prefer. Pick out the best brush for the job, remembering that large, radial shapes smooth and straighten, and thin ones curl. Draw hair ends on to the brush, winding them round as you would spaghetti on a fork. Secure the hair firmly in the centre – not spilling on the handle or over the tip. As you blow above and below each section, ensure that the hot airstream moves with the hair, not against it.

For a basic bob dry, centre-part the hair and clip back the sides and top. Starting on the underneath sections of the nape, take small, manageable layers to achieve maximum fullness and volume in the area. Proceed up the back of the head towards the crown, unfurling strands of hair carefully as you go. Steadily dry strands from top to tip before rewinding and eventually removing the brush. Take down the side sections and dry, drawing the hair out from the head in a backward direction. Tackle the front last, curling the hair under. When individual strands prove difficult to shape, leave the brush to cool inside the curl and the roundness should then remain. To prevent charging hair with static electricity when working near to the scalp, direct heat off it at an angle.

Styling pointers

- To give the roots and the parting body and lift, raise the hair off the head at a 90-degree angle and dry beneath the strands.
- For general volume dry from below.
- To obtain a feathery effect blow dry the hair against the cuticle, pointing towards the root.
- Remember that hairspray reduces flyaway, flatness and frizz.
- To bring polish to a sleek, straight style, brush the top layers softly and blow dry from above.
- Curly and wavy hair won't straighten perfectly, not even temporarily.
- For a non-bumpy fringe, flick hair back, dry and brush forward.

When the Heat Is On

- Around the hairline direct the hot air current in the opposite direction to that in which you want the hair to lie eventually.
- *Scrunch drying* means that hair is towel-dried first then blow dried while the stylist holds small clumps of hair in one hand and squeezes.
- *Rub drying* means the palm of the hand is rubbed along the scalp in circular movements to create height and a back-combed effect.
- *Grip drying* means the hair is twisted in sections and tightly held while being blow dried.
- *Blow drying* is the means to an end, the icing on the cake.

We assume your hair is beautifully cut and in tiptop condition so that drying adds the final touch of perfection! Heavier hair weights derive the greatest benefit from blow drying. The extra natural body bulk allows for increased styling variation. Finer textures can also be enhanced but no blow dry shape in fine hair holds very well or for very long. And if a style demands more than average drying time, ask yourself if the end product is worth the trouble. A perm may be a better alternative. It would definitely be more enduring, probably kinder to the hair in this instance and be cheaper in the long run. Remember that healthy hair is also susceptible to heat over-kill. Heat-damaged hair can be nursed back to health with regular treatments of reconditioning shampoo and conditioning cream.

The rules for blow drying apply to all hot styling equipment – hot brushes, tongs, curling wands or crimping irons. A word of caution: stick to straightforward models until you master wrist control; the rotating barrel and swivel cord varieties can be aggravating for the novice. Most brushes and rods have the following as standard features: an indicator light to signal the tool is switched on and heating; a release button for interchanging heads; thermostatic control and a rest-stand.

Steam models come out tops for giving the gentle treatment. Extra dry, brittle and processed heads of hair go through torture with plain dry heat. For all hair types non-stick, coated tongs are a good idea because they eliminate the chance of burnt remains adhering to the metal.

Clip sections not being styled out of the way and commence work at the back of the head, gradually proceeding upwards. Wind hair round the styling barrel, tucking in the ends, and roll up to just short of the roots. It's tricky to do at first but you soon get the knack. Comb curls and waves out once hair cools, for maximum effect.

Restrict hot styling sessions to a couple of times a week at most and less if you use a crimping iron – the waffle-like tool that gives a corrugated, ripple effect and is extremely drying. Basically, the less you use any drying or styling instrument, the fewer occasions there will be to create damage.

10 A Permanent Alternative

If the word 'perm' conjures up a long, shaggy, 'coupe sauvage' or a head full of tight 'afro' curls, throw away the notion. That was yesterday, when fabricated 'natural' curls were still a novelty and the hows and wherefores sometimes not fully understood. Modern perms are in another class altogether. They are an exercise in professional technical expertise resulting in diverse styles, so soft and *real*, that half the time you wouldn't know the hair was permed.

Since its early days the field of perming has enjoyed seemingly endless success in the research and discovery of new work methods. Techniques and products alike have undergone rapid development and when you consider the number of advancements there have been in ways of winding alone, the future appears very bright indeed.

It was about the turn of this century that the first perms came into being and they have been making more and more waves ever since! Perming has contributed a major part to the revolutionising of the art of hairdressing, experiencing phenomenal progress itself in the process, and stimulating original thinking and inventiveness throughout the industry. For the client, the biggest benefit has undoubtedly been the expansion of styling options: whole head perming; curl on the ends only; partial perming on top and sides; texturising – where different hair thicknesses are produced by skilful varying of rod sizes at intervals over the entire head. The choice is yours.

Of course, this wasn't always the case. The original perms on the scene before the Second World War, were heat induced and the resulting waves, strong and very artificial-looking, almost 'channelled' into the hair. In those days, long, deep, wavy styles were the fashion and to cater to the trend, perms were spiral wound from roots to ends to produce the sought-after effect. The chemical solution put on had to be heated to near boiling point before it would alter the hair's structure. The whole affair must have been quite an ordeal for client and permer.

The arrival in the 1940s of the cold wave perm (which operates at normal temperatures) allowed a more popular kind of winding called croquignole to supersede spiral waving. This method introduced the concept of winding in the reverse sense, from ends to roots, and parallel to the head's shape. The ensuing curliness was far superior to anything that had previously evolved. However, in comparison to today's freer outlines, these perms looked manufactured, stiff and totally lacking in subtlety. Not until the 1960s could hairdressers reproduce waves and curls 'Nature's' way. The birth of the 'Greek Goddess' look by Vidal Sassoon meant that instead of hair being finger-waved, pin-curled or roller-set in place and dried under a hood, it was left to dry naturally for the first time. To all appearances, the perm was completely unstructured and even a little messy if not actually wild. Hot on the trail followed the 'Coupe Sauvage' – wash 'n' wear and 'drip-dry'. Suddenly a market, once primarily patronised by older women, opened wide to attract younger women, and men of all ages.

While the bonuses of a perm are obvious to everyone, think long and hard about the steps involved before you book one. Permanent means just that. You cannot wash those curls out! For the permanent chemical reaction to take place on the internal structure, perming power has to break through to the cortex layer of the hair shaft. To speed penetration, hair is shampooed and towel-dried before the solution is applied, which is normally during the winding of the hair on to perm rods. Washing the hair beforehand swells the hair slightly and causes the cuticle scales sheltering the cortex to rise gently, allowing the perm lotion faster entry. The lotion alters the keratin in the hair and breaks down the sulphur bonds linking the cortex cells, so loosening up the individual fibres. When stretched over curlers in this chemically altered shape, hair adopts the perm rod's outline. More lotion is applied once the wind is complete and then the head left to 'develop', that is, the time the perm takes to fix in the new shape. (Hair in poor condition may find this transformation stage hard to take and become crumbly on the inside.)

Development time is a variable factor. Not even a thoroughly knowledgeable hairdresser can predict the precise moment the perm will have sufficiently processed the hair for the intended effect to be achieved. The choice of product strength differs according to what the hairdresser considers is necessary for the job and the hair's texture and condition heavily influence how soon the 'stop point' is reached. For example, a tinted head perms at a faster rate than virgin hair (that is, non-processed) because it's considerably more porous after any chemical treatment. Like waiting for a pot to boil, the perm's progress needs constant watching. Your hairdresser will test a sample rod on a part of the head notorious for developing slowest – the nape, for example. Over-processing, that's when the perm's left on too long, causes frizzing. But if the developing time is cut short, you may feel you never had a perm at all because it will fall out almost immediately.

In any event, once the decision to perm is taken, if the new outline is to be permanently retained the changed links must be reformed into their new shape by means of a second chemical reaction. This is done by using a neutraliser and there's no leeway for a change of heart after this stage. Hair is rinsed while still round the rods, excess water blotted off, and the neutraliser applied. After a few minutes the rods are removed and more neutraliser put on before a final rinse and conditioning. How does it work? The neutraliser contains an oxidising agent which is effectively responsible for closing up the broken sulphur bonds and leaving behind the design line. The degree of movement established is there for everyone to see. The curl/wave strength depends on the type of perm used.

The traditional-style perm lotions still presently in use, but to a much reduced extent, can be relied upon to give a crisp curl when the occasion demands. Don't be put off by the rather awful smell which originates from their ammonia content. The newer, softer brands (based on sodium bisulphate) are sweet smelling in contrast, with recognisable fruity fragrances like apple and strawberry. Unfortunately, the other significant difference between the two, the softer curl result, means that the effect falls quicker. This 'body' perm as it's called, gives movement and volume without actually creating defined curls or waves. It introduces body of a casual, gentle variety, effective for a shorter time than it would take to grow out. However, although a body perm principally produces bounce and vitality, the hair has nevertheless been chemically changed and will remain so until completely replaced by new growth. The alternative full permanent wave product, naturally enough, creates distinctive movement and a firm base for greater manageability plus added life. The finish is crisp and best suited to controlled styling. With either type, good conditioning is a must even if the perm has a built-in conditioning element, as many do these days.

Perming techniques

Successful perming depends on many things besides good timing and products employed. Although the over-riding factor might logically seem to be the style you want, the dimensions of the rods (length and diameter) and their overall shape, plus the hair's length and texture, predetermine the eventual outcome. This is where the permer's skill, in correctly choosing the most appropriate options, comes into play. They have to be right first time. For example, there is a maximum amount of hair that should be wound around the curlers. Too many strands give an uneven result. There's the

A Permanent Alternative

line of the cut – executed before or after perming. There's the winding direction – from a multitude of possibilities one will stand out for the hair's requirements in terms of texture and thickness. Technique names you'll hear your hairdresser talk about include directional; spiral; rik-rak; stack; double decker and weave.

Directional winding

Perm rods are placed the same way rollers are for a set. Hair is sectioned and wound on to rods laid in the direction the style line will follow.

Spiral winding

Modern-day spirals create distinctive waving through the entire head of hair, with strands wound from the roots to the ends and positioned horizontally on the scalp. Hair is wound round specially long curlers designed for the purpose and resembling sugar candy twists. This method is most suitable for longer hair because of the quantity of hair required for 'spiralling', and will produce a romantic pre-Raphaelite look or very chunky, natural waves.

Rik-rak winding

Basically another name for the procedure above.

Stack winding

The design pattern is built out from the top of the head, leaving hair straighter all round the top and very curly on the ends. The perm rods are gradually constructed into a halo effect and secured by long pins.

Short hair permed upwards and forward and coloured in the front

Double-decker winding on large rods gives full, chunky curl

A graduated perm with a wispy front and full back

Body-permed to achieve gentle volume for bounce

A combination of weave and double-decker winding achieves a fabulous full line

A Permanent Alternative

Double decker winding

This wind is pretty much like the bus of the same name, with two layers, one on top of the other. Again ideal for longer hair, two rods of different sizes are used to give varied ripple movement with firm curl on the ends. The larger of the two rods is wound from the ends to the halfway point on each hair section, the smaller rod is tucked under the first and both are rolled into one hair portion and secured on the head.

Weave winding

This technique also requires two rod sizes and gives super volume. Hair sections are wound on to curlers in fine strands, with the top strands going on to the larger rod and the underneath ones on to the smaller. In this way, a mixture of loose and tighter curls is woven through.

Since people discovered you don't need conventional tools to perm, all sorts of other ideas have been explored. Small plaits and twists made over the head can be permed and then taken out. Long pins like knitting needles make super soft waves, while triangular rods are great for sharp lines, and some ingenious permers even use coathangers and chopsticks. The possibilities are mind-boggling. If your hair will play the game, the only other thing you need is an adventurous permer ready to have a go.

Professional versus home perming

With the technicalities involved in even the most basic perm, you can understand why we are totally against home perming of any description. For a perm to score on all the previously mentioned levels, we believe the professional permer is the only person with sufficient know-how. The burnt scalps and frizzed ends previously associated with perms are now thankfully a thing of the past. Most salon staff take pride in their art, and the interest of the trade as a body to become 100 per cent competent in the new ways has been tremendous. The advances made in the last decade alone compare very favourably with the parallel growth of computer technology, and hairdressers are privy to new information hot off the presses thanks to trade journalists. Salon staff can participate in training seminars somewhere locally on at least one night each week, and manufacturers, no longer content simply to sell their goods, organise professional teach-ins for hairdressers using their products to ensure clients benefit from them in the way they were intended. However, given that despite the high standards existing, a few hairdressers still make blunders, what chance do you, the novice, have of coming out a winner? Operating outside the professional's environment, at best you can expect to produce uneven curl or patchy movement and at worst you'll cause extreme dryness and hair breakage. Not only does perming your own hair take in all the planning aspects already listed, but you must be prepared for any mishap. A product is as good as the hands it gets into and while a home colouring mess may be rectifiable, a perm isn't.

But why bother with home perming anyway? If it's a question of saving money, do you? For less than the price of a home perm pack, you can have an excellent perm put in at student training evenings in salons, hairdressing schools and technical colleges. All students are closely supervised. No one in authority can afford a disaster so in some ways you are almost assured of closer attention than during a normal appointment.

Is a salon perm really beyond your reach? A professional perm can appear expensive but when you consider how long it lasts and divide the price by the number of weeks on that basis, it really isn't. A top salon charges for time – the time it takes to examine the scalp and talk about your hair's textures and previous treatments (a re-perm job requires an entirely different approach from virgin hair, and processed hair of any sort demands a weaker solution be used), and the time required to work out the mechanics of rod winding corresponding to the hair's length and the movement direction of the style to follow. To give clients the best possible service, many

A Permanent Alternative

salons now employ permanent waving specialists who work with the stylist to determine the line of the finished look. That's what you're paying for. Should you come across a salon with prices far below everyone else's, ask yourself why. It could well be that staff put on a lot less lotion than is needed, or use too few rods, so their perms have sadly reduced holding power. There will be some reason that works to your disadvantage, you can be sure of that.

The pros and cons of perming

If you're advised that your hair's not in good enough condition to stand up to a perm, for instance, it might be dry and split, do listen. Don't insist that the perm be done, or rush off to another salon. While a second place may agree to do it, you can't be sure they're not just less scrupulous than the first salon, and only after your money. To resolve any doubts, ask for a strand of hair to be 'test-curled' at the outset. This way you and the permer can assess the likely result.

Perming bleached or highlighted hair is completely out unless you don't mind looking like a haystack; and a perm on newly tinted hair is a bad idea because the perm lotion lifts the colour from the tint and so alters the shade. It's best to give hair and scalp time to recuperate between processes. The perm should come first with a rest period before tinting of a week. Split ends (or with colour build-up on them), irritated scalp, overly dry or brittle hair, and extra fine textures which may or may not 'take', all deserve to be carefully examined before being put forward as perm candidates.

Post-perm maintenance among clients is notoriously poor. Why a new perm owner should be any different from a new car buyer, who automatically expects to have to service his pride and joy to keep it in good order, is a mystery. A perm's lifespan is likewise considerably prolonged with a little loving care and conditioning protection. Avoid pulling and stretching curls with brush or blower, and never shampoo immediately after a perm because it continues to be chemically active for several days. Shampoo could well provoke an adverse scalp reaction. By popular consensus the biggest bonus of a perm is its 'shake back into shape' ability after a wash or swim, but each time round more of the natural oils seep away. Remember to replace them on the ends and the lengths permed, leaving conditioner on a reasonable time prior to rinsing.

Perm grows out as hair grows, which is about half an inch per month. The only alternative way to get rid of those curls is to cut them off. Fine, soft hair shows fallout quickest, and flattens down where the permanent wave lift was originally compensating for the naturally lacking body and bounce. Aim to retain the style shape permed in so that you've a good looking design line as the perm grows out. Unfortunately, some hairdressers fail to produce designs that can be recut into another great shape after a few months.

Sooner or later, every perm has its day and becomes sad, droopy and out of condition. Another perm is clearly required but that's impossible because the old one has not completely gone, and the hair hasn't grown sufficiently to permit the perm ends to be cut off. The answer is to have only the regrowth permed now, and to protect the previously processed sections from the chemicals – much like recolouring is done. This way you maintain a decent design line from day one, up to perhaps half a dozen perms later, provided care is taken throughout this highly specialised process. Permed shoulder length hair keeps its curls for as long as four to six months but short hair requires a second perm after three months simply because the perm is just giving lift and bounce or chunky movement. Six-weekly salon visits allow your stylist to keep an eye on the perm and for regular general care between visits, shampoo to encourage lift or dampen with a water spray and use the fingers to redefine the shape.

A permanent solution

Perms are a godsend for limp, lank, fine and flyaway hair, and a blessing for all other types too. Cost conscious clients want a

A Permanent Alternative

style that will last and a perm does. Full, bouncy curls and ripples of waves don't have to be what a perm is all about. Many of the sharper outlines benefit enormously by having the edge taken off their harshness with soft subtle perming. Styling variety has increased twofold now that cuts are executed after perming as frequently as before curlers are wound. Practical solutions to hair problems are on the up and up too. Perming slightly thickens each individual hair strand by causing them to swell and helps greasy locks stay oil free longer by creating wave movement which slows down the flow.

The first-time perm client should be very explicit as to type of movement wanted. We've seen a lot of clients disappointed because they conveyed the wrong message and hair has been made too wavy or not curly enough, or the entire directional movement has been irregular. Bear in mind that the perm starts growing out almost as soon as you leave the salon and during the weeks that follow you'll face a few difficult moments unless you're advised on aftercare. Your hairdresser may suggest a weekly visit which you probably will think is overdoing it, but if the perm is a drip-dry model, that's a jolly good idea at the beginning to assist you with shaping. Don't have hair cut too short just before a perm because a certain length is required to 'take up the curl'.

Perming to straighten hair might sound a bit of a contradiction but it can be done and is. Excessively curly hair can be permed as normal using the same chemicals but performing the method in reverse and combing through for a straight or, if very curly, a wavy texture. If you think there can be problems with permanent waving, just wait until you hear about straightening. In many instances hair softens or processes sooner than expected but not necessarily for the eye to see right away. Because straightening is most often carried out on fragile negroid hair, breakage frequently occurs unless professionally done with an excellent product. We talk about this at greater length later, but for now, suffice it to say that we totally oppose the domestic use of straightening chemicals. We've seen too many horrendous examples of home attempts to be able to condone it. Damage from straightening is far more extensive than anything seen with perming for the simple reason that the process is executed from the scalp up. And please, please, never have a perm then change your mind because you hate being a curly top and request a straightening session soon after. You'd be crazy even to consider the idea. It's too soon and too risky. Wait for a couple of weeks at the minimum, until the curl settles down. You may just find those curls grow on you but if that doesn't happen, they'll at least have become looser in the intervening period. Further encourage decurling by setting hair on the biggest rollers you possess, leaving them in until hair's completely dry to avoid frizzing. Or 'iron out' your troubles by straightening with a styling brush during blow drying. Perhaps ask your stylist to cut your hair in a style that'll remove the bulk of the curl and continue to do so at each cutting session.

The modern perm can give Mother Nature-type curls, if that's what you want, or if a subtler touch is more to your liking, will produce just a hint of movement. Perming is one of the few salon techniques that allows such a degree of leeway and is a big success story as far as the public is concerned because of the freedom it endows. Hair still looks naturally a part of you and takes the least amount of looking after, even when you lead an extremely active life. But there's no doubt about it, perming is rough on hair, especially to the dry and fine varieties. It's nice to know most manufacturer's ranges carry lines specifically for already weakened locks, like bleached and tinted, but conditioner is nonetheless paramount in combating hair's now increased vulnerability.

Whatever way you look at it, however, the benefits of perming far outweigh the inconveniences. Just ask our clients!

11 *Colour Is What Counts*

Imagine a world without colour, everything in shades of black and white. You can't really, can you? Colour fills our field of vision everywhere we look – in our food, our clothes; at work, at play; blues, yellows, pinks, browns, greens . . . colours are all around. Colour is a vital part of our existence and without it our lives would be immeasurably poorer. No more blue skies, yellow sun, green meadows. No more stopping at red traffic lights or not parking on yellow lines either! Nature provided the base colours and we have expanded upon them so that we now actually communicate with each other in colour terms of reference.

Red equates with danger; black is reserved for sombre or sophisticated occasions. We say someone is 'green with envy' when they are jealous, or in a 'black mood' when angry. We have expectations of others according to the colours they wear. Dull, drab clothes translates to dull, drab person. Greying hair means 'she must be getting on in years', and blonde hair spells 'dumb'. We talk and think colour all the time but, for the most part, are unaware of the dominant role colour has come to play in our lives. We respond on cue to colour coding, colour commands and the influence colour brings to bear on our feelings about people and situations. So you can see the power hairdressing colourants could come to exercise upon your life. Contriving different colour environments for yourself can free you from the image you perceive others to have of you and dislike, or give you the cosmetic freedom to create, in part, another you. As with make-up, practice makes perfect and only over time will you and your colourist come to maximise on colour's creative powers.

The trouble is, too many people are still convinced that a colour transformation is a drastic measure to be taken only when age or premature greying makes it a necessity. Hair to them is a head covering and they accept their natural shade as the one that was meant to suit them. This is rarely true in actuality, and with a little help from colourants almost every head of hair could be made the beauty asset that Nature never really intended it to be.

Another area of confusion relates to the types of product available and the effect they have on hair. Modern colourants have been so improved in quality, range and capacity that they no longer resemble the products of even ten years ago. Techniques of application have similarly progressed so that colour lights, tips and flashes are as feasible as all-over tinting. That's the beauty of today's colour – you can have as much or as little as you want.

Choosing a new colour is like picking out a make-up. The shade should complement the face's skin tones and eye colouring. While some colours enhance youthfulness, harsher ones are ageing. Take a look at the salon's colour charts in the special tinting area, if adapted to give daylight conditions, or step outside the door for just a moment. A colour's true tone is hard to assess under artificial lighting and you may find yourself with a result that's a few shades darker than you expected, if you make a decision based entirely on looking at it indoors. Also, have your colourist test the one you like on a few strands of hair, to help you both decide on its suitability. Avoid dramatic leaps from dark to light shades. Gradual change is less of a shock for you and your friends, to say nothing of being cheaper. Extremes from brown to blonde, for example, show up regrowth in a matter of days and upkeep

Tortoise shell highlights.

All over colour using Wella burgundy.

Deep blue silk dye hair accessories.

Zigzag lights in three tones to create texture.

Mahogany vegetable semi-permanent colour plus protein moisturiser treatment.

Avantgarde colour used to complement the total look.

Highlights with added blonde tips to accentuate movement.

Strips lights giving a shimmering effect.

costs can bite into your budget substantially. Don't be dictated to by fashions. Some hairdressers talk about the 'Year of the Blondes' one year, and the 'Year of the Brunettes' the next. Ignore all that and wear what suits you. Here's a simplified guide to colours that complement skin tones.

Black hair – burgundy, wine and auburn lights break up the density of the colour. *Medium light brown* – golden chestnut has a nice softening effect. *Natural auburn* – coppery tones match best. *Reddish/blonde ginger* – golden lights. *Mousy/light ash blonde* – light tinted blonde lights. The main objective of artificial colouring is to enhance the hair's own tones so we always work with part of the natural pigment as our base.

You'll also hear your colourist refer to the 'tones' and 'shades' in hair. The way to differentiate between the two is to think of hair types as coming in ten *shades* from black to white. The *tones* are the entire colour spectrum – neutral, red, gold, ash, etc. For example, rich brown, chestnut brown and auburn brown are all the same shade – brown – but have different tones.

Temporary rinses

The mildest and least permanent form of colour is the temporary water rinse, so called because it washes out in the very next shampoo. As a means of introduction to the world of colour, there's none better. A rinse only coats the outside of the hair, it doesn't penetrate to the cortex, so you can experiment in the knowledge that the colour will come out with the next wash.

Temporary rinses add depth, warmth and vivacity to natural colour but don't produce a startling change. For example, naturally grey hair often has a slight yellow tinge to it caused by smoke fumes and air pollution. A silver rinse will take away the yellow, if the client wants her natural shade back, while a slate grey will give extra depth and a darker effect.

Water rinses work on blondes best of all. When hair is tinted blonde a tremendous amount of the natural pigmentation is removed. At the same time, the colour tint carries ashening agents through the hair which eliminate any brassy tones. These agents fade rapidly, within a couple of weeks, or sooner if the client goes into the sun or uses strong detergents that strip the colour out. By using a silver or platinum rinse to replace the ashening agents in between retouching sessions, the tinted blonde is assured of a consistent colouration.

These rinses can be applied to all virgin hair, that is, non-processed, but they really work best on light-coloured hair. Available in shampoo, setting lotion or post-shampoo form, they can be used at home without any problem unless a bad colour match is made; like vibrant red added to bleached platinum blonde hair. Because bleached hair is very porous, the rinse would stain the sponge-like cortex and the colour take a great length of time to come out.

Principal advantages of the rinse are its cheapness and the fact that you can have a colour try-out, see what you think, and then make a decision for a more permanent technique. But for some, the colour result would probably look too false, and the product does come off on bed linen and shirts, and if the shade is dark and you're caught in the rain, you'd feel rather foolish. Also, while a rinse gives hair more body, the coating does have an unflattering, dulling effect.

Semi-permanent rinses

These are based on synthetic vegetable colours and fill the market demand for colour lasting longer than the temporary rinse but not as permanent as a tint. They add colour to hair, creating richer and sometimes darker tones that enhance the natural shade. For example, a semi-permanent on dark brown hair will bring out the same chestnut tones that the sun acting on the natural pigment in summer produces. And marmalade blonde hair, which is dull and lacking in reddish lights when there's no sun about, perks up fast with the application of a golden chestnut or soft honey blonde 'semi'. But semi-permanents cannot lighten hair colour.

Colour Is What Counts

A semi-permanent rinse penetrates the transparent hair cuticle and deposits colour deeper within the cuticle layer without the activation of an oxidant. The cream or liquid product is applied to shampooed, slightly towel dried hair and left on for 15 to 30 minutes. The colour change is perceptible but not strikingly obvious, fades gradually and normally lasts four to six washes, but on very porous hair, which accepts colour more readily, the time may be extended to eight washes. There is no real regrowth problem unless the hairdresser has completely mismatched the semi-permanent colour with the rest of the individual's natural colouring attributes.

Not only is this one of our own favourite colouring techniques but our clients love it too! They are able to experiment with colour without having to make the major commitment to a complete colour change that the permanent tint requires. The product's versatility can be taken advantage of by a whole range of clients: the 'try anything once' client who takes no risks following this route; the 'refresher' seeker who wants a little sparkle in her life gets just that; the 'I don't want to go grey yet' lady doesn't have to – a semi-permanent provides the means of disguising a few stray grey hairs here and there; the regular tint client who has the semi-permanent at her disposal, for application between root retouching visits, if her colour starts to fade too quickly.

The biggest advantages are in cost and condition terms. The process is relatively inexpensive and with no regrowth to worry about, you only return to the salon as you wish – not for colour maintenance. Some semi-permanents contain added conditioning elements but it doesn't matter if they don't because most are acid-based and so encourage the cuticle to lie flat, reflect light and shine beautifully.

Tints

Permanent tinting is among the most versatile of all the colouring techniques. The range of tones available, both for lightening and darkening hair, is enormous and colour coverage of hair is total. The result is so convincing that hairdressers are sometimes hard pressed to tell if the colour is natural or not! Even grey can be completely masked.

How tints work: equal amounts of hydrogen peroxide and tint are blended to activate the colour. When applied to dry hair (middle, ends, then roots) the mix penetrates to the cortex where the natural colour molecules are located. Deposits of tint combine with your own pigment to produce a brand new shade. Developing time takes 20 minutes plus, but can be speeded up by heat. All the tint must be washed from the hair afterwards or the colour will begin to change on contact with the atmosphere.

Tinting is harmless if carried out correctly. Today's products are so advanced and reliable that if anything really goes wrong, the fault lies with no one but the tinter. Of course *you* must be 100 per cent honest in relating your colour 'history' – what sort and how long ago – but in the final analysis, your tinter's judgement is what counts most. He or she needs to estimate how much pigment is already in the hair to decide if your hair is in a good enough condition to withstand the chemical processing involved. Regardless of whether or not your hair has been tinted before (the same holds for all colouring except vegetable dyeing), we recommend you come into the salon 48 hours prior to your appointment, for a patch test. This is performed to check for possible allergic reaction by your skin to the contents of the tint to be used (for example, peroxide or chemicals). All we do is dab a little tint on the skin behind your ear or on the inside of your elbow. If sensitive, the skin will become blotchy and puff up within 24 hours. Report back to the salon at once if you have any such symptoms. They will advise you on what your next move should be and may suggest another product for you to try but for now, tinting is out.

We cannot impress upon you too much the importance of this pre-tinting step. Colouring products are mild and don't usually harm anyone but in those rare instances when a client is really allergic to something in them, the skin and face, and eyelids in particular, swell up. In severe cases, the scalp forms pus-filled spots and there may even be a higher temperature that takes up to a week to subside.

Colour Is What Counts

The permanency of tint means the colour has to *grow* out and that the 100 per cent alteration will endure for several months. Look at the salon's manufacturers' colour swatches in daylight and choose a colour you can live with. Permanent tint can make your hair colour as much as six shades lighter than the original; for example, from a dark brown to auburn or golden chestnut. However, the ideal amount is one to two shades lighter. We personally won't tint darker than the natural colour because the finished result is flat, minus natural highlights and looks false. Some younger clients request a complete colour change and while we're happy to do this we will not alter the *shade* very much. You see, everyone has at least three to four different tones in their hair and if you turn hair darker the colour becomes a matt, solid mass and looks dyed, which is contrary to what we work for. By tinting one or two shades lighter instead, the natural tones surface to the same degree, reproducing the natural colour in the hair. Regrowth is less obvious too.

What about perming tinted hair or vice versa? While not in favour of either one, occasionally it is in order if hair is healthy and bouncy or the client has had regular deep conditioning treatments to reinforce her hair in between the processes. Never, ever, have hair permed and tinted on the same day. A perm should always come first because if done in reverse, its acidic content will lift the tint to a lighter shade than desired. Leave, at the minimum, a period of a week between processes because the perm's chemicals continue to act on the hair for several days afterwards, and an application of tint would simply pull the perm out. What you must appreciate is that doubly processed hair needs an awful lot of aftercare to merely survive. Tinted lights and semi-permanents are gentler means of introducing colour sheen to a perm. If, however, you want a tint every six weeks and a perm about twice a year, you must leave an interval of two weeks after perming before re-tinting, and have regular moisturising sessions. Tinted-and-permed hair ends are extra porous and layers of colour make them look heavy and unattractive.

Tints are marvellous for grey hair. When grey begins to show we use semi-permanents to cover up, but a stronger approach is advisable with 30 to 40 per cent grey. The first thing to take into account is the amount of pigmentation already lost from the skin. To tint hair back to its own shade would drain all colour from the face and leave behind a solid mass of the tones the client had twenty years previously and which would now contrast starkly with their features. We prefer to tint hair one or two shades lighter and to tone the grey streaks down to a shade lighter than the new colour on the hair bulk. In this way, the natural colour attains slightly lighter overtones where the grey hair was originally. One reason why tinted hair with some natural grey shows up the grey first as the colour begins to fade, is that today's tints are designed to be so gentle to hair that they don't have the staying power of the old-fashioned, stronger varieties. The natural colour has only been made one or two shades lighter so has less fading capacity than the grey, which is perhaps four to six shades darker and will, therefore, fade dramatically. To overcome this problem, a semi-permanent of the same colour should be applied to the whole head about three weeks after tinting. The base shade won't become any darker but the grey will vanish.

Tinted hair can become dull after a few weeks for two reasons. Firstly, non-complementary products have been used at home. Tint is alkaline-based and opens the hair's cuticle. Shampoo used subsequently must have a pH factor of 5.5 and conditioner be acidic within a pH range of 3 to 3.2 to close the cuticle. This is why it's imperative you understand more about hair's home needs. Ask your salon for advice on home product use. Secondly, tinting is somewhat comparable to medicine consumption in that your body builds up a resistance to one drug and you have to change to another. Hair is exactly the same and after time a new formula might be required to arrive at former results.

Because a tint fades rapidly, you will require a regrowth treatment approximately every four weeks, though frequency is controlled by the degree of contrast between the old and the new shades. The fresh tint goes on to the regrowth first, for twenty minutes, and is then combed through to the ends and left for 10 minutes. Touching up the roots is the time-consuming and wearisome aspect of permanent tinting but is ultimately less expensive than the initial visit.

Bleaching

Whole head bleaching is something we haven't done for the last fifteen years except for special effect, at a beauty editor's request, for a magazine feature. For example, we once bleached a negress's hair blonde and added a bright colour flash to produce an extraordinarily effective style photograph. But in all other instances, bleaching is taboo. To bleach hair even just two shades lighter kills natural lights and creates a flat, solid, brassy blonde look which is unflattering to the most attractive of women, to say nothing of the impact it has on the condition. As we said earlier, colours which are a few shades lighter than the natural pigmentation suit the complexion. People who favour bleaching want hair six or seven shades lighter. If you are unable to go blonde without tinting and have to resort to bleaching, your natural colour must be so dark that your skin contains yellow pigment and your eyes are dark brown – in which case, you'd look terrible as a blonde.

Bleach, which is largely ammonia-based, is mixed with an oxidising agent – hydrogen peroxide – to give a combination that removes hair's natural pigment but does not replace the colour. Only after the correct base shade is obtained, is blonde toner added. When regrowth is treated every month, the new bleach has to meet the line where the old bleach stopped before. However, even if this is successfully achieved, the bleach's swelling action causes it to overlap on to previously processed portions. By the time the dark roots have lifted to yellow or white, the overlap area is being lightened a second time around, and further weakened. A blue/white stage is eventually reached, where the hair's structure breaks down to a stretchy jelly. No self-respecting hairdresser would deliberately allow this to happen but it does, inevitably, after years of such treatment. And because all-over bleaching is essentially destructive, restructural treatments must play an integral role in the battle to lower the hair's now increased porosity. Occasionally, improvement can only come with cutting.

On top of all this, to revert back to your own colour at a later date will take a lot of hard work and be extremely costly. Quite frankly, bleaching's not worth the trouble. The way the colouring market has developed, bleaching is very much a 'golden oldie' and easily omitted from the bulk of the new techniques. A tint will give a vastly superior degree of subtle movement through hair.

Lit up in lights

Lights of any description, call them what you will – tips, highlights, tinted lights, lowlights, rim lights, halo highlights, painted streaks, shimmers, tortoiseshell – are all for one principle purpose, to put more 'light' into hair. Lights are the nicest way imaginable of accentuating the four to five different tones in each head of hair. Cutting and perming can accomplish wonderful illusions of movement but never create a less than solid appearance. By contrast, lights shimmer and shine in rippling patterns of colour. The golden rule is: no colour is to be introduced into hair that is not already present, and for any that is introduced to be near to the natural shade. Over-soft highlights can seem greyish.

Highlights

Highlights (streaks) are fabulous for clients who want a gentle transformation without monthly salon visits, but should be reserved for mousy blondes to ensure a natural looking effect. Highlighting is one of the oldest colouring techniques known to man and involves bleaching isolated strands of hair. Legend has it that the women of Phoenicia were among the first to make use of this method. They are reputed to have pulled hair strands through small openings in the straw hats they wore while working in the fields, and applied a lemon juice or chamomile mix to them. Both solutions are hair lighteners, which the sun activated.

But it was not until after the Second World War that highlighting really came into its own as a fashion tool, when the Parisiennes initiated the silver foil method (explained below). A more time-saving but less precise method, employing a 'plastic' cap, was introduced by the Americans in the 1960s. Both techniques remain popular today. We personally recommend the foil method on the grounds that the cap is uncouth, un-professional and pretty painful. Once again, you get what you pay for. Lights by the cap method cost about £12, take over an hour to do and frequently come out uneven. The finer foil sectioning takes several hours and is far more expensive, but is executed with such precision, in a procedure not unlike erecting a brick wall, figuratively speaking, that you are definitely assured of value for money in terms of final results. Like the builder, we work in zig-zag, horizontal lines to ensure that the hair will never fall into vertical colour stripes, regardless of styling. This leaves behind highlights which almost perfectly imitate the sun-induced varieties. Here's how the two diverse methods work.

The cap method

A polythene cap is fitted over the head and individual hair strands are drawn through purpose designed holes with a crochet hook. The bleach/hydrogen peroxide mix is spread on to even-sized sections and left for 20 to 30 minutes depending on the base colour and the desired result. Once hair is shampooed and rinsed, but before the cap is removed, an oxidizing toner is applied to the highlights. Then the toner is washed off, the cap removed and the hair shampooed as normal.

But this method prevents the colourist getting close enough to the parting to highlight correctly. Consequently, strand sections often vary in thickness, causing irregular lighting. Friends will probably ask where you had your highlights done because they show! Retouching is fraught with problems. There's a tendency for colour to be re-applied to the whole head every time, making it practically impossible to avoid damage to the hair, and causing the ends to be repeatedly treated. The hair then generally lightens all over and begins to look more like a tinted blonde.

Silver foil weave method

While the oxidation process works in the same way, application is totally different. Hair is pinned into neat sections and fine strands, about ten at a time usually, are separated out. With a strip of aluminium foil underneath the sectioned strands, bleach is brushed on to the hair, then the foil is folded into a compact parcel to isolate hair not intended for processing. There can be as many as 100 such packages on the head at the finish, depending on the fineness of highlights requested. The whole affair takes several hours for an entire head highlight, and performed thus, is the most expensive of salon services. But the results are worth it and compliments will flow.

You need to return after three months for a colour retouch to the front hairline, natural parting and crown. The roots over the complete head are usually tackled three months after that, with colour applied solely to the regrowth. The ends are not repeatedly reprocessed as they are flicked out from the parcel packs.

Tinted lights

Ideal for lifting brown and mousy shades out of the doldrums; tint rather than bleach is woven through the hair for lights. They serve as a fantastic pick-me-up for dull, winter hair.

Lowlights

Best on light brown or blonde hair; very mild bleach produces lights softer than tinted lights.

Rim lights

This technique works well on a mid-length style with full fringe. The colour rim undulates delicately through the hair but in no way resembles a demarcation line which could hint at old tint growing out.

Tips

Nice for curly styles; the principle for application is the same as for ordinary highlighting except that larger sections of hair are taken in between weaving the highlights to avoid the risk of creating one main tone.

Painted lights

Fine streaks of mild bleach are painted through the front of previously tinted hair to add illusionary movement.

Halo highlighting

As the name suggests, highlights produce a halo effect which is most becoming and pronounced on layered hair.

Tortoiseshell lights

This is a lovely way of introducing more than one colour into hair. Beneficial to brown shades particularly, as many as three to four colours are put through the hair in alternating, individual strips. For example, red one strand, hazel the next, then honey blonde and blonde.

Shimmering

Only about 20 per cent of the hair is coloured, with a combination of permanent and temporary shades which grow out naturally. The method is ingenious in as much as there's little regrowth and the colours can be added to or detracted from, altered or retouched. The effect is a shimmer of colour dancing like rays of sunlight.

The thing to remember about highlights and tinted lights generally is that any toner applied to the entire head must be peroxide-free to avoid bleaching the natural colour and creating a regrowth problem.

Vegetable colours

Henna had been the mainstay colourant of cultures in Africa and Asia in days long before Cleopatra discovered the dye's merits. Nowadays, as well as being used by Eastern women as a bridal decoration (when intricate patterns are drawn on the palms of the hands and soles of the feet as part of the marriage ceremony), henna is popular for its conditioning and dyeing attributes. Currently, in the West, henna is most commonly used for hair colouring and has taken on a predominant role in the natural products market. Henna comes from the *Lawsonia* plant, which smells like roses and originates from Africa and Asia. Moroccan and Indian hennas have reasonably good conditioning properties and an orange/red pigment. Persian (Iranian) henna is the finest in the world, both for conditioning and pigment yield per leaf. Its colour is far redder and richer than any other henna.

There are only two authentic hennas, neutral and red. Of the others on sale many are compound substances containing metallic salts which are fine, so long as you steer clear of tints and perms, otherwise, the ensuing chemical reaction causes hair to disintegrate. Here's what happens to a tint: light brown hair tinted medium chestnut and then covered with henna lightens all over, because the tint is stripped out, and exhibits chestnut-coloured regrowth.

Henna is planted annually and in the first year neutral henna only is produced. The colourless leaves are dried and crushed to a powder after deveining. Mixed to a paste with hot water, the product is applied after shampooing and towel-drying hair and left on for 30 minutes under a hot towel to speed developing. Neutral henna gives body and a marvellous shine.

Into the second year, the henna leaf starts to yield red dye and when put on hair, stains the cuticle. Like other herbal dyes, henna colours more slowly and with less penetrating depth than artificial tints; but it also fades less quickly. Some say the results are unpredictable and that's true, in the hands of inexperienced operators. On the first occasion, red henna application is total – middle and ends to begin with, and when the colour is half-developed, the roots – taking about an hour in total. The lighter the base, the redder the finish; the darker the natural shade, the more coppery or chestnut the result.

Because henna develops with heat, whole head treatment every time is out as repeatedly expanding hair like this makes it brittle. Root retouching at four to six week intervals suffices, and for extra shine and to revive hair's bulk colour, an appropriate semi-permanent free of peroxide and heat requirements is a good idea. Red henna works best on dark heads – black, brown and brunette; not on grey or blonde, which turn carroty.

Although we have said there are only two true hennas, a black henna can be made up that is quite pure in nature. Combining red henna with indigo, which is bluish in tone, gives a brownish/black shade known as black Persian henna, which is utilised like the other two.

Herbal dyes, while not as widely used as henna, have vast potential for extending their popularity. (Note that herbal and vegetable in this context are interchangeable terms as far as hairdressers are concerned.) It is mainly advocates of natural products who take advantage of these dyes' capacities but the health aspect needs to be kept in perspective. There is little point in insisting on herbal colours for your hair's health and then smoking 100 cigarettes a day. Nor are their conditioning qualities superior to those of synthetic colours.

Chamomile

This most commonly used of lightening herbs is prepared like tea. Once brewed and left to cool, the infused mix is strained to obtain the lightening solution and applied cold, as a final rinse after normal shampooing and conditioning. Used on very mousy blonde hair regularly after every wash, successfully produces a slight lightening effect. Hair is guaranteed to lighten with the herb, when used over a period of time, but not after just one application. Chamomile has only natural ingredients to lift colour so works best with lighter tones, for example, blonde and mousy, where pigmentation is so much weaker and hair finer, but not brown. Good results are eventually evident. Health shops sell chamomile.

Rhubarb Root

Common old garden rhubarb roots are the strongest of all herbal lighteners. They have to be boiled for over an hour to elicit the lightening agents, and the longer they're left, the higher the yield. Used as a cold, final rinse on dark brown to light brown bases, the colour change is not instant but can be speeded up by a spell out in the sun just after application, to give golden highlights there and then. Because rhubarb root has a certain amount of yellow content, blonde and white hair can't be treated or they'll adopt a yellowish hue.

Saffron

This is probably the most expensive herb to buy. It resembles root ginger in appearance but when cut open is bright yellow inside. Saffron is available from some food shops in standard powder form. Boiling extracts an incredibly vivid yellow dye which colours instantaneously. The solution can be diluted according to the strength required, and used as a water rinse. Saffron performs well on blonde hair whether natural, tinted or bleached, giving a yellow flush to it. You have possibly seen saffron rice served in an Indian restaurant and remarked upon the colour. The shade imparted to hair would be a diluted version of that.

Marigold

This simple little flower, found readily in the United Kingdom, is mixed with boiling water, cooled and used exactly like saffron as a final rinse. The slightly weaker yellow tinge softens lighter bases.

Sage

Found in almost any local supermarket, this herb is infused like chamomile and has darkening properties, but several applications are necessary before even a slight difference in colour is apparent. Generally used on grey hair, to take away whiter elements and to tone the colour down to a more silvery slate.

Most tree barks and nuts yield a tan of sorts and though primarily used for tanning hide, a couple can be prepared as colourants.

Cascarilla

A tree whose bark yields an extremely strong black dye, which is extracted in an infusion and used as a last rinse. Grey hair can be covered nicely but the tones from the bark do vary relative to its own colour depth.

Walnuts and hazelnuts

After roasting the shells only until almost burnt, they are ground to a powder and mixed with water to form a paste. The slightly dull colour that appears after treatment can be brushed or washed out. For a more permanent dye the shells should be boiled for two hours plus. Frequent application brings darker results.

A treatment of our own based on avocados and called

Avocado Wax Treatment, has fantastic cleansing powers and provides super sheen and body.

Herbal colours represent a viable product alternative to the synthetic ranges, but beware if you use them at home because the stronger ones like saffron can be unreliable. The controlling factor is the hair's porosity; the higher it is, the more colour soaked up. Basically, if you know what you're doing you should have little problem.

Colours of the spectrum

When people think of wilder colouring the more bizarre, ugly and tasteless examples on view in the high street come instantly to mind. The 'way-out' combinations can be horrific eyesores and, not surprisingly, this type of colouring has earned an unjustifiably bad name. Employed with style and artistry, hair can be brightened in a fun way for special occasions or just because you feel like being unconventional for a while.

Our own Spectrum colours are based on silk dyes. For best results with the lighter colours, blue, pink, canary yellow and green, the section to be coloured has first to be bleached quite light. The Spectrum colour is reapplied every four to six weeks but bleaching only takes place once. With the stronger shades like burgundy, deep chestnut and wine, we add tinted lights to hair first and then the Spectrum colour, which mustn't touch the roots. The colours are safe for home use but we'd rather you came into the salon first time round for instruction tips. How permanent are they? Shade used and hair condition are the influencing elements but they do have a general staining effect and a deeper red obviously lasts longer than a pastel blue. To return to your own shade means the Spectrum shade will have to be stripped from the hair and that section toned to your natural one. Over the ensuing weeks you'll notice tremendous fading on the previously bleached section and it will seem as though another regrowth is coming through. The bleached section eventually grows out.

Powder sprays

Although they only coat the hair, and are great fun for someone who is not a colour user in the normal way but wants a mind-blowing party look, some can leave metallic deposits on the inner cuticle. Again, contact with permed or tinted hair could cause a chemical reaction. However, the metal bases themselves are unharmful and the powder brushes out.

Clients often ask us questions about colouring. Here are the answers to some that pop up regularly.

What is colour build-up?

When too much pigment seeps into the cortex build-up occurs. This happens with semi-permanents which colour the outer cuticle. If a client returns for further sessions sooner than they should for example, every other week, the colour depth will increase in strength until the hair is badly stained. Excess colour causes no damage but is extremely hard to remove.

If a hairdresser says a head has acquired terrible 'build-up', he means that every time the hair has been coloured, dye application has been total, so that colour layers are now evident. For example, if for eight sessions dark brown hair is tinted all over light chestnut, the colour becomes no richer but the pigment and chemical build-up is difficult to get out. The worst case comes with continuous whole head henna applications. The colour deepens and the constant heating causes the hair to deteriorate.

How often is a complete colour change wise?

You know what suits you best and if you're a mousy blonde it's unlikely you'll go black one month and deep auburn the next. In any case, extreme colour changes are inadvisable and undesirable because artificial pigment lifts less easily than natural. A slight change, from auburn to copper, for example, is best.

What technique is best for fine/coarse hair?

Very fine, baby hair should not be highlighted, tinted or permed. The hydrogen peroxide required to activate these processes opens the hair's cuticle and in the course of a few treatments, strips hair of its oils and causes structural breakdown. Semi-permanents are far more gentle and definitely the most suitable.

Coarse hair is, as you know, slightly porous and ideal for absorbing semi-permanent colour type too. A permanent tint would serve equally well, but why use a harsh treatment when a milder one will do? Coarse hair is brittle and a tint's harsher action may make it an unwise choice.

What colour works best on Negroid hair?

You would expect this hair, of all the brown base shades, to be the hardest to bleach to blonde. In fact, the pigment lifts faster than its European or Oriental counterparts do. A head can be bleached to pale lemon, from point to root, in one hour! The bleach is so fast acting because of the hair's porous nature. However, we'd never put bleach on an entire head. Negroid hair is too brittle and can't withstand the trauma of such heavy processing.

Tinted lights can be dismissed too because the hair is so frizzy. No matter how large the sections coloured, the lights disappear into the mass of curls when dry, despite looking very effective when wet. We normally colour the hair's top layer one or two shades lighter. Brushed on finely over the top eighth of an inch, the shade can be enhanced when developed and rinsed by adding a semi-permanent wine, burgundy or deep auburn to promote a pretty, soft halo line.

If hair is bleached or tinted blonde, can highlights be added?

Yes, at a price. If your hair was toned back to its natural shade and highlights bleached in, the fee would be considerable and the results attractive – for three to four weeks – if you're very lucky. The toned-down colour fades quickly and also acts as a toner on the highlights, making them duller. A reverse situation now exists. The toned hair has lightened and the highlights have darkened, so the total is muddy and dull. You would return to us four or five weeks later unhappy that the highlights had disappeared and the rest had faded. This happens earliest with bleached blondes, but highlights on tinted blondes still mingle with the bulk colour to look murky. The only way of achieving highlights successfully is for us to tint around the parting the same, now lighter shade, to match up the blonde. To begin with, we would tint the natural parting and crown for three to four months. By then the underneath sections are beginning to grow out, but not to a noticeable extent. We change your style and part your hair where there's regrowth of about two inches and then introduce lights of the same colour as the ends, into the roots. The effect isn't great by any means, and we won't be happy for six to eight months more, but this is the best changeover procedure we know of which enables you to remain blonde throughout. At the halfway conversion point the highlights suddenly overtake the tint and appear as lights with strong sunbleach on the ends, which should be cut off.

How long will hair take to recover from bad colouring?

The answer to this question depends entirely on the initial degree of damage. If over-processing makes hair feel like wire wool after washing,(before conditioning), or like slime when dry, there's little to do but to have it cut off and await healthy new growth. If switching colours too often has caused bands of colour like dark brown, auburn and orange to surface, we can offer some help. Hair would be gently stripped of the colour and we'd work towards making the rest of the hair the same as the darkest colour. We would attempt to lift the dark brown band to an auburny tone and bring the other tones as near to that as feasible, to obtain an even result. The bands cannot be stripped out individually as this would create a lighter band of colour on either side of the dark one.

With all-over blonding, is there any way to have a subtle effect?

We can tint the existing parting the same colour it has been for several months and then relocate it at the side or middle where there's regrowth. Then we would introduce lights of the same colour as the tint or, if the hair's colour is too light, perhaps put in tinted lights and a semi-permanent rinse over the rest to tone the lighter colour to blend with the highlights. This way there's less of a 'full' blonde feel.

What do we think about home colouring?

Not much. People often tint or bleach only the spots they can see, for example, around the face, and not evenly through the hair. Colour is patchy, and if applied all over every time, the roots become a different shade and the hair like wire wool. Why give yourself such unnecessary headaches when a professional can colour your hair better for you?

Remember too, whatever you take out of hair must go back in. Hair should be in good condition prior to any colouring, and after processing needs at least one deep treatment a month to rebuild and remoisturise.

12 Child's Play

A vast number of parents approach the issue of their children's hair care rather matter-of-factly. Getting them out to school in the morning, neat and tidy, is a job in itself and a quick, daily brush and twice-weekly shampoo is about all that they can manage. But hair care is as much a learning habit for children as most other things are at this stage and the maxim, 'the sooner you learn, the better', applies to a greater extent to them. Peculiarly though, while the young are scolded to brush their teeth after meals and to wash grubby hands, etc., hair care is made to take a back seat when it shouldn't.

The early years

Ideally, children should be able to manipulate their own hair by about the age of eight or nine. Up to that point, mums and dads, the task of setting a good example with a routine of your own, is yours. Encourage imitation early on. You can't start too young. But try not to bring any elements of fear into the activity because bad experiences in early days can put children off hairdressing for a long time into the future. Like too many dentists, too many hairdressers have memories of younger clients being dragged into their salons screaming blue murder because they imagined some terrible horror awaited them inside.

Starting with shampoo nights, consider your child's finer feelings and cast your own mind back to those early wash days. Most of us hated them. If shampoo didn't seep into our eyes making them sting like crazy, lukewarm rinse water trickled down the back of our necks, wetting a shirt or dress collar. Right? Well childhood experiences don't change that much, and shampoo time is still a nightmare in some households, with exasperated mums grappling a fighting youngster who is determined to wriggle away. Forward planning is the key solution. Arrange the family timetable so that you don't have other urgent matters to attend to immediately afterwards, and so feel pressurised into bringing the shampoo session to a rapid conclusion. Try not to leave washing to the minute before a family outing. If even the slightest thing goes wrong, bedlam will ensue.

Introduce a sense of fun into the proceedings and perhaps even create a salon-like atmosphere. Sit junior in a chair backed up to the wash basin and gently tilt his or her head backwards into it, so that water won't run on to the face. Test the water temperature with your elbow, which is more sensitive than tougher-skinned hands, and use a spray hose for more efficient heat control and manoeuvrability. Keep a towel nearby in case of panic stations with water in the eyes. Use a non-irritant shampoo specially formulated for children to eliminate stinging; adult shampoos are harsh for infants and can cause skin rashes. And please don't wash hair at bathtime. Apart from being unhygienic, any bubble bath suds in the water may contain mild detergent which could be too strong for fine baby hair.

Massage shampoo in with the fingertips and never scrub the scalp roughly. Where babies are concerned, remove scurf caking the scalp in infancy by softening the scales with olive oil. Leave it to soak in a while and then gently prise the scurf loose with cotton wool. As baby grows scurf should disappear but if not, consult a paediatrician. Baby's hair needs washing like everyone else's but not with a forceful shower spray.

Child's Play

Cradling the baby in your arms, scoop water by hand on to its head from a wash bowl or basin.

Conditioners certainly aren't necessary for babies, and usually not for toddlers either, because their hair is the healthiest it will ever be and remains so for much of childhood. A slightly older child with dry scalp may need a mild anti-dandruff shampoo and a moisturising rinse but, by and large, children's hair stays in tiptop condition quite naturally. The only exception might be long hair. An anti-tangle rinse helps to reduce knotting and avoids the 'mummy, mummy, don't pull' entreaties.

Many parents wonder when to schedule the first salon visit. The right time varies as much as children do but whatever happens, please don't force your child into a cut. Wait until they seem ready. You can encourage things along by taking them to the salon on one of your appointments. This way they become familiar with the surroundings and the staff. Most salons cater for younger visitors with comics and toys, and a very few have instituted crêches.

How you handle the first cut is important. It would be nice if the decision to go ahead could be a joint one and the chosen style a compromise of tastes, your own and the child's. Opt for a gentle transformation and if you can just think for a moment about how fastidious you are with your own hair, imagine the sensations this first visit awakens in your child – they will be nervous, apprehensive and fidgety. They'll look to you for confidence, reassurance and approval so remember to offer compliments afterwards. Presuming the first session passes off without incident, establish a pattern of reasonably regular cutting or at least trimming, as required.

Mums often leave daughters' hair long in the early years because it's 'so pretty'. Those same shiny, flowing locks can also cause your child much aggravation. Right up to the point when Joshua's daughter, Maria, had her very long, thick hair cut short at the age of ten, she was in agony every time she combed out tangles, and had a continual uphill battle keeping strands out of her eyes at play. Plaiting, braiding or a ponytail brought some control but the weight she was carrying around occasionally gave her headaches. Parents who for sentimental reasons keep a child's hair long beyond the day she herself wants it aren't being terribly fair or practical.

Nor are parents who have a daughter's hair cut 'pudding basin' style (so short she's never allowed to look feminine) or who force their son to sport ringlets until school age. You know how children tease each other, and the last thing you should want to make them is a subject of ridicule with their peers.

The notion that children's hairdressing should be cheaper than adults' is a bit of an old-fashioned one. Quite truthfully, as much skill, time and patience is devoted to the former as to the latter, for the simple reason the little rascals hardly sit still long enough to allow you to get a pair of scissors near their heads. Add to this the extra persuasion tactics called for and the equally intricate angles sometimes cut, and you begin to see why prices are probably reasonable. We only mention cutting because we don't consider perms or tints are for children. Mid-teens is soon enough. They're only young once and there are plenty of years ahead for them to experiment with the range of professional hairdressing skills.

At home, encourage participation at every stage. Keep separate combs and brushes for junior family members and organise cleaning shifts. Let girls pick out clasps and ribbons to wear and coach them in the use of a dryer. For roller setting fans, big, soft foam models are safe and fun. Sooner or later someone will have a go at clipping their own hair. We've all had our moments and it's not the end of the world. Hair is easily straightened but keep scissors out of reach to avoid a repeat performance. To prevent nasty rubber bands going round ponytails and pigtails, keep a handy supply of coated bands.

The day all parents dread is the one when a child comes home itching with head-lice. People panic and suffer pangs of guilt thinking, because lice are associated with dirt, that they too must have failed on cleanliness. While it's true that lice breed in unsanitary conditions, it's also a fact that once one child in a class is infected, the others follow. Children are in such close contact in the course of a day that there's almost no way to prevent an epidemic. Lice are transmitted when hair

Child's Play

strands rub together or on to clothes. They lay greyish-coloured eggs (nits) close to the hair shaft and mainly at the nape, in small, ring-like patches. Getting rid of them requires painstaking combing with a special steel comb and applications of Malathion or Carbaryl solutions, two effective eradicaters obtainable from a chemist. Sprinkle DDT on pillows and clothes to disinfect them but don't expect the powder by itself to kill the pests because most modern strains seem to have developed a resistance to the chemical. If the problem persists in long hair for weeks, the quickest way to get rid of the pests is to have the hair cut. You probably won't hear too much dissent from your child because constant itching is very uncomfortable. Head-lice infestation is undoubtedly unpleasant but keep calm, and after a series of treatments the whole nasty business should soon be over.

Teen times

At puberty, the children who previously had to be cajoled into washing their hair suddenly become so hair conscious that you hardly recognise them. *He* combs his hair anywhere, anytime and preens himself in front of the mirror; *she* plays around with rollers and dryer putting in waves and curls. Half the fun of being a teenager is experimenting with fashions and knowing you can get away with it. Hairdressers appreciate the role young people play in keeping the hairdressing industry revitalised, energised and exciting. The 'Teddy boy' and 'Punk' trends both emerged from teenage circles to become high-street salon styles. Unfortunately, this phase of self-expression sometimes pushes parents to the limits of their patience, but if everybody can just stay on friendly terms the period of strain will pass painlessly enough.

In their late teens some boys exhibit the first signs of baldness and worry terribly. Joshua began to lose his hair at 18 and was virtually bald on top by 26, so knows the feeling. If you are losing your hair we honestly believe the best you can do is to learn to live with the thinning. If you can accept early on that you're going bald, your lack of self-conciousness will cause others to notice the problem much less.

However, if you become really depressed as the thinning progresses, you may want to ask your stylist to help you choose a good quality, natural-looking hairpiece that will give lasting service. Do this as soon as the thinning takes hold so you don't find yourself in the position where you've hardly any hair one day and a full head the next. That's not exactly the discreet approach, though we'd encourage you not to view your condition as something you need to hide. You really don't.

13 On Longer Lines

Long hair is a truly wonderful asset when it is maintained in top condition. Men fantasise about women with long hair and other women watch with envy when long-haired women swish by and every glance follows them. Unfortunately, too many owners of long locks keep the length even when it ceases to suit them or is no longer healthy enough to warrant it. With some it is purely a matter of vanity; with others, a question of fear. They are frightened to have it cut because long is the only way they've ever worn it, and even if the hair is straggly and split, because it doesn't grow well, the familiar is easier than the unknown. Unfortunately, there does come a time in a woman's life when age catches up with her, when long hair should not be worn loose to the waistline but cut mid-length or worn up. The pity is, some women don't recognise when this stage is reached and from the back their image suggests a youthfulness that's far from the truth. Who wants that? Others who won't face up to the truth are those whose hair type is naturally not suited to grow beyond a certain length on all or parts of the head (the hairline, for example). Decide what's really best for *your* hair.

Conditioning care

Long hair, because it is around so much longer, obviously suffers more from wear and tear than short hair. The ends,

Hair up into French pleat and drawn forward to form a quiff roll

Simple but chic moussed chignon

An inverted plait for all occasions

Long hair twisted and tied with rags for casual control

which may be several years old, have borne the stress of brushing, shampooing and detangling over all that time. Needless to say, the hair is fragile and so conditioning is vitally important, and get rid of split ends, which only continue to work up the shaft until they are cut off. After shampooing, remove knots with the fingers and a wide-toothed comb, working from the tips up. Brush with a natural bristle brush to give the hair polish, clean it and carry its own conditioner, sebum, down the shaft. Never, ever, brush hair when it is wet as this causes it to weaken and eventually break.

Try to dry naturally whenever you can and keep away from hot irons and other electrical styling equipment which dry the hair out further. Use good quality tools, hair ornaments with smooth edges and covered elastic bands to avoid snagging hair.

The styling options

Perming and colouring are possible but only at the salon, not at home. Setting doesn't make sense for hair longer than mid-length because the strands are too heavy and bulky for the waves to stay in. However, the new foam shapers and spiral rods, that are often as long as the hair itself, successfully produce gentle waves or curls. The big plus for long-hair wearers is in the areas of styling and dressing. The options are numerous.

The *chignon* is just a ponytail twisted into place and secured with a band on the top or at the nape. Tackle the sides by twisting and attaching first one side and then the other with pins. A *bun* or a *roll* can be quickly made by using a padded roll (a fibre-filled and netted shape that will give the volume and the outline design you require). Just position the pad where you want the bun or roll and cover with the hair. The *roll and plait* calls for hair to be parted across the crown from ear to ear and the back put in a ponytail. The front is taken off to the side in a small plait, tucked into the ponytail clip and the whole lot twisted into a chignon. The *French pleat* involves hair being directed to one side and then folded in towards the middle to pleat and pin. The *French twist* is hair gathered at the back, twisted, pinned and the ends tucked in at the top on the crown. The *top knot* is all the hair gathered towards the crown, wound with an elastic band and twirled into a knot positioned over the band and secured. *Plaits* African-style have become a hit with Western women particularly since Bo Derek wore her trendsetting style in the film *10*. The more ethnic-orientated scalp plaits and dreadlocks are, however, probably only for the truly converted because they are expensive to have done and take hours to put in.

Hair accessories plus fashion rings (also padded but in ring shapes), ribbons, scarves and false pieces, can all be utilised to give an added dimension to long-hair styling. But probably *the* indispensable tool is the decorative comb which, unfortunately, a lot of people do not know how to use properly. Flatten the hair with it first then insert the comb in the opposite direction to the one the hair has to go in, reversing it so that it grips.

To cut or not?

Deciding whether or not to cut the hair can put some women in a real dilemma. They may have worn it long since childhood and feel apprehensive about what such a change would entail. After all, it's always hard to imagine any style change on your own familiar face so give a thought to the agonies someone with really long hair goes through when thinking about having it all cut off. If you have long hair in top condition and wearing it long suits you, great. But if the extra time it takes to look after is more than you want to give and you've decided to become a 'shorty', then to make the move less drastic, go for a mid-length transformation first. From long to very short right away is too dramatic for even the most adventurous among you. And be selective in your salon choice because, at the moment of reckoning, you'll feel comforted if you know there's a friendly face or two around who understand your predicament.

Hair Matters

Once you have made up your mind don't let anyone dissuade you, and at the salon make things easier for yourself by not watching the scissors at work and the hair ends falling to the floor. Just tell yourself it is all a bit of an adventure and whatever the end result, it will be a pleasant surprise. Be ready to pay for a good cut. After the relative low cost of the trims you have had in the past, the price of a smart, new cut may come as a shock. But if you think that there is no going back once your hair is gone then surely it is in your own best interest to ensure the new look is a stunner.

If you've decided to keep hair long then you are obviously happy with how you are presently coping and unlikely to be a regular salon client because you have your own little ways of handling your hair. But don't forget to have ends shaped or your hair won't stay healthy, glossy, sexy and admired by all around.

14 *Dressing Up*

Bridal beauty

If there is one day in your life when you want to be a knock-out, it's your wedding day. But there is so much to organise and never enough time to plan it all, that sometimes things are more of a chore than a pleasure. The key to a successful countdown to the big day is planning. Start by preparing hair weeks in advance, getting rid of split ends, scurf, old perm and colour, and growing hair longer if it is to be put up. Enlist your stylist's help in finding the right style and take along the headdress you'll wear, if any, plus a sketch of the dress, to create a total picture. Witnesses and bridesmaids should be in on the arrangements if they are to wear complementary styles and colours. Book a series of appointments for major members of the wedding party.

Colour and perm changes should be done several weeks before and don't go for a new look just because it seems to be the thing brides do. You may feel more comfortable dressing up your everyday style with a few fresh flowers or pretty combs. For short hair, a good cut may be all that's required while longer hair could need several deep conditioning treatments to be on top form. The thing to think about is how you will wear your hair on the honeymoon, or even back in the

Naturally curly hair enhanced with vegetable colour for that special day

Sophisticated yet saucy hat and hair combination

office, if it has been shaped into a high fashion style that's not too practical. Is it worth the bother?

Set aside time for a couple of dress rehearsals when you can experiment with make-up and your outfit to achieve the 'total look'. Allow, too, for the extra stress bringing on spots or adding to existing hair problems. Keep calm and reflect on the romanticism of the occasion. Here's the golden opportunity to indulge in some olde-worlde elegance by accessorising with ribbons, flowers, combs or ornate Alice bands; to wear hair in small chignons or simple coils and plaits. Make the most of this very special day and enjoy yourself.

Hair accessories and hats

No fashion-conscious lady can have missed the important role hair accessories now play in the overall fashion picture. Pulled out once upon a time for Christmas parties only, combs, clasps, feathers, pins and thongs are now used whenever the notion strikes. Perhaps you are in extrovert mood and want to impress with an eye-catching fantasy design, or simply to brighten up your everyday style. Whatever the reason, there's bound to be something to fit the bill among the large variety of accessories now on sale in department stores and salons. There are plain and patterned slides, grips, clasps, feathers, chignon pins, *diamanté* ornaments and bands to choose from. For extra special occasions there are shell, sequinned and pearlised decorations, or super satin and velvet bows to make a splash. For a completely co-ordinating accessory wardrobe look out for the sets of hairbands, grips, bracelets, earrings and brooches that some companies make. And, of course, there are often practical reasons for using ponytail clips, banana grips and headbands to hold hair in place or back off the face, such as on the sportsfield.

Hairdressers used to feel that hats had no place in hair-styling because they cover up the very thing the stylists work to enhance, but not any more. Hats are big news since the Princess of Wales demonstrated how dashing they can look working *with* hair. Most popular are the soft felts and natural straw boaters which, perched to one side of a head of soft curls, are positively head-turning. Hair in plaits and bunches sit prettily under a straw saucer-shape or bowler while the really adventurous can opt for a crownless brim where hair is pulled through the centre and arranged around the frame in loose bunches. To facilitate wearing hair up with a hat, some hats have an elastic strap underneath to help hold them on, or you can roll hair into a bun and secure your hat in place with a hat pin pushed through the centre.

Fantasy styling

Put a little sparkle into your life with glitter sprays and gels. An evening style with a touch of shine sprayed on strategic spots will have all eyes on you as soon as you walk into the room. When spraying, wrap a cape round your shoulders to keep particles off your clothes. Fantasy sprays with a fluorescent effect can be shown to most flattering advantage applied through a cut-out stencil pattern in the shape of your choice, like a flower or a bow. To achieve a more overall controlled effect however, go for gels. These are ordinary gels with gold or silver metallic deposits in them, or coloured in one or several fashion shades like pink, blue and green. Most wash or brush out relatively easily so you can have great fun seeing what artistic heights you can reach in the knowledge that none of it is permanent. However, if you pop along to your salon for a perm or a tint soon after use, do let your hairdresser know if you have recently used a glitter spray or gel so as to prevent the chance of a chemical reaction.

Glossary

allergy   an adverse skin reaction to a chemical(s) or other substance(s).
alopecia   a scalp disorder leading to baldness.
arrector pili   the name of the hair muscle.
back-combed   hair is combed down towards the root.
blunt cut   hair ends are cut straight across with little curl left in the ends.
bob cut   as blunt cut.
caucasian   European-type hair.
colour build-up   when too much pigment seeps into the hair's cortex.
cortex   the inside layer of the hair which is made up of fibre-like cells.
cowlick   an area on the hairline where hair springs up in a loop.
cuticle   the flat, overlapping, transparent scales that make up the protective outside shell of the hair shaft.
dandruff   large grey or yellow oily flakes on the scalp.
desensitised   hair that has been chemically processed too often and becomes fragile within its internal structure.
double crown   the head has two crowns, one in the usual position and one just below.

finger waves literally waves pressed into wet hair with the fingers.
follicle the slanting passageway below the scalp from which hair emerges.
gel a setting agent.
graduated cut a cut which produces hair shorter in the nape, progressing to longer at the crown.
henna a vegetable dye from the *Lawsonia* plant that comes in neutral and red forms.
highlights light blonde streaks put through hair.
hydrogen peroxide the usual oxidising agent used for lightening and tinting.
keratin the protein that constitutes the bulk of hair.
mousse an aerosol-type setting agent.
over-processing when permed or tinted hair is left to 'develop' longer than is recommended.
papilla the base from which hair grows.
patch test before all-over tinting or perming, a little product is put on the skin to test for possible allergic reaction.
pH refers to the acid or alkaline level of a substance and is estimated on a scale of 0–14. Below 7 denotes acidity and over 7, alkalinity.
pin curls hair is worked in fine strands wound around the finger.
processing the activity that results when hair is chemically treated.
regrowth new growth that comes through after hair has been tinted.
restructurant a product designed to help hair's internal problems by working on the cross-linked keratin chains which make up the hair.
retouch application of tint to new hair growth.
scrunch dried hair held in small clumps is squeezed with one hand while being blow-dried with the other.
sebum the skin's own lubricating oil.
section to divide off or to part hair.
semi-permanent a rinse of synthetic colours which lasts through four to six washes.
static when an electric charge is set up in hair through brushing, etc.

Glossary

straightening   when hair is straightened by chemical agents.
tapered   hair is cut with scissors slightly open and run along the shaft to taper off.
temporary rinse   a water rinse that washes out with the next shampoo.
texturised   different hair thicknesses produced by varying rod sizes when placing for perming.
tint   permanent colourant.
trichologist   a hair and scalp specialist.
virgin hair   natural hair that has never been chemically altered by colouring, perming, bleaching, etc.
widow's peak   the fine point of hair that grows backwards from the middle of the forehead.
winding   wrapping hair round a curler or other styling shape.

Index

accessories 50, 132, 138, 139, 143
acidic balance 22, 23, 28, 111, 113, 114
age, effect on hair 5, 6, 19. *See also* grey hair
allergy 41, 112
alopecia 42
American Indian type hair 14
amino acids 3
anaemia 42
androgen 42
ash blonde 6, 109
auburn hair 109
avocados 123–4

babies 130–1
backcombing 46, 74, 76
baldness 41–2, 133–4
bath, washing hair in 25, 130
black hair 109
bleached hair: colouring of 110, 126, 127; perming of 103; effect of sun on 50
bleaching 115–16; effects of 5, 38, 42; at home 46; of negroid hair 126; for highlights 116–19
blonde hair 6, 10, 109, 110, 118, 122

blonde tinting 110, 126–8
blow drying 16, 75, 81, 86, 89–91
bob 16, 62, 65, 68
'body' 11, 12, 32, 90
body perm 76, 96,99
body shapes 59, 60, 63–5
breakage of hair 11, 17, 27. *See also* ends
brown hair 6, 10, 109, 110, 118, 119, 122. *See also* mousy
brushes, quality and cleanliness of 24, 37, 44, 46, 132; hot 90–2, 106
brushing 24, 37, 42, 138
buns 67, 138, 144

cap method of highlighting 117
cascarilla 123
Caucasian hair types 10, 12, 14, 16
central heating 29, 38
chamomile 121–2
cheekbones 65–7
chemicals, damage by 5, 12, 17, 46–7, 105–6
chignons 15, 63–5, 67, 68, 70, 136, 138
children's hair 129–34

Index

chin   65, 66, 69
Chinese hair   14
cleanliness: of salon equipment   54. *See also* brushes; combs
clothes, relationship to style   59, 62
coarse hair   10, 11, 14, 15, 19, 126
cold wave perm   94
colour   107–27; natural colour   5–6, 10, 14, 36; colouring techniques   11, 14, 15, 108; effects of colouring   28, 50, 125; choice of colour and technique   108–9, 111, 113, 138; home colouring   46–7. *See also* auburn; bleaching; blonde; brown; flashes; grey; highlighting; mousy; perms; pigments; redheads; regrowth; rinses; 'shades' and 'tones'; spectrum; sprays; tinting; tips; vegetable colours; white hair
comb, decorative   139
combs, quality and cleanliness of   37, 44–6, 132
commuting   24, 29
complaints   57–8
condition of hair   31–42
conditioners   11, 12, 22, 31–6, 38, 49, 74, 96, 104, 106, 113
contraceptive pill   41, 42
cortex   3–5, 10, 14–15, 94, 95, 109, 110, 112–15, 125
cosmetics. *See* make-up
cost. *See* prices
cowlick   8, 18, 70
crimping irons   91, 92
croquignole   94
crown, double   13
curls, perms for   91, 93, 94, 96–8, 101, 102, 104–6. *See also* pin-curls
curly hair   5, 10, 11, 13, 15, 17, 62, 74, 119. *See also* frizzy hair
cuticle   4–6, 31, 32, 37–8, 47, 95, 111, 114, 121, 125, 126
cutting: as basis of style   61, 71–4; methods of   72–4; blunt   73, 74; graduated   65, 73, 74; layering   73; tapering   73; fine hair   11, 62; long hair   139–40; children's hair   131–2; for perms   97, 104–6; remedial   47; price of   140; home cutting   44, 132. *See also* bob; short cuts

damage to hair   5, 11–13, 17, 44, 46–7, 91–2, 105–6, 127. *See also* breakage; ends; restructurants
dandruff   28, 36, 37, 41, 131
detergents   23
development times   95, 112
diet   6, 19, 28, 36–40, 42
diplomas and certificates   53
dirt   5, 21, 24, 26, 28, 29, 36
double chin   69
double-decker winding   100, 101
dreadlocks   139
drugs, medicinal   36, 38, 41, 42
dry hair and scalp   28, 31, 34, 36–8
dryers, use of   11, 37, 38, 42, 44, 86, 89
drying techniques   37, 77, 79, 83–92
dull hair   12, 23, 31, 47, 110, 114

ears sticking out   69
elastic bands   132, 138
elasticity of hair   5, 8, 27, 35, 37
ends: cutting of   72, 73; damage to   11, 13, 28, 37, 38, 44, 47, 74, 135, 138, 140
eyes, position of   66

face structure   56, 59–61, 63–71. *See also* cheekbones; chin; ears; eyes; forehead; glasses; jawline; neck; nose
fantasy sprays   144
fashion   56, 61, 109

Index

fine hair: as type 5, 10, 11–14, 16; control of by cut or perm 62, 104; colour treatment of 126; blow drying of 91
finger drying 84, 87
finger waving 75, 79, 81
flashes, colour 17, 108
'flyaway' hair 11, 12, 28, 31, 37, 90, 104
foil method of highlighting 117, 118
follicles 2, 6–8, 10, 12, 17, 39
forehead 65–9
fringe 18, 66–70, 90, 118
frizzy hair 15, 17, 62, 90, 106, 126
French pleat 136, 138–9
French twist 139

gel 74, 82, 87
glasses 70–1
'glassy' hair 36
glitter sprays 82, 144
greasy hair 6, 13–14, 25, 28, 32, 34, 36, 37
grey hair 6, 108, 109, 111–14, 121, 123
grip drying 91
growth of hair 2–3, 6–7, 12, 18–19, 39, 62, 63, 72, 73

hair: structure and function 2–8; type 9–19. *See also* condition of; cortex; cuticle; follicle; growth of; length of; loss of; medulla; papilla; scalp; sebaceous glands; strength of; texture; virgin hair
hairdresser–client relationship 54–8, 60, 71
hairline, irregular 13, 15, 18, 70
hairspray 11, 12, 37, 46, 74, 82, 90
halo highlights 116, 119
hard and soft water 36–8
hats 29, 37, 142–4
hazelnuts 123
health, effect on hair 6, 19, 41, 42
height. *See* body shapes

henna 35, 120–1, 125
herbal dyes. *See* vegetable colouring
highlights 103, 116–20, 126, 127
holiday care 47–50
home care of hair 42–50
home cutting 44, 132
home perming 46–7, 101–3
home straightening 17
hormonal activity 6, 10, 36, 41–2
hydrogen peroxide 112, 115, 117, 118, 120, 126

iodine 39

Japanese hair types 14, 15
jawline 61, 66, 67

lacquer 14, 38, 46, 76
lamp drying 84
lanolin 28, 32
lather 21, 22, 24–7
layered hair 73, 119
length of hair and styling 5, 27, 63–70, 72, 73. *See also* growth of hair; long hair; short cuts
lice 132–3
'lift' 63, 64, 66, 90
limp hair 12, 36, 62, 104
long hair 36, 135–40
loss of hair 7–8, 40–2, 133–4
lotions: perming 95, 96, 103; setting 32, 46, 76, 78, 79, 81, 88; colour rinses in 110; for frizz control 17
lowlights 116, 118

make-up 59, 62, 67, 108
marigold 123
massage 3, 28, 37, 38, 42
medicated shampoo 22, 41
Mediterranean type hair 15
medium texture 10, 14, 62
medulla 4, 5
melanin 5, 47
menopause 7–8, 19, 42
menstruation 19, 36, 40
minerals 39, 40

150

Index

moisture content of hair 6, 15, 28, 36, 37, 113
mousse 80, 82, 85, 86
mousy hair 109, 116, 118, 122, 125

neck, shape of 59, 64, 65, 69
Negroid hair 10, 14, 16–18, 105, 115, 126
nits 132–3
normal condition and texture 10, 11, 36
nose, big 61, 70

oestrogen 7–8, 40, 42
off-the-face styles 63, 65–8, 70
oils, natural 23, 25, 36, 104. *See also* sebaceous glands; lanolin
Oriental type hair 10, 11, 14–16

painted lights 116, 119
papilla 2, 3, 7
parting 8, 66, 67, 69, 90
patch test 112
perms 93–106; effect on structure of hair 5, 28, 38, 74, 94, 95, 105; over-use of 42; effect of health on 40–1; effect of sun on 49; effect of colouring and tinting on 103, 113, 120, 125, 126; for fine hair 11–13, 62, 74; for long hair 138; for different face shapes 64, 65, 67, 68, 70; for Oriental hair 14–15; home perming 46–7, 101–3. *See also* body perm; curls; winding
peroxide. *See* hydrogen peroxide
pigment 5, 6, 19, 110, 112, 114, 115, 120, 125
pigtails 132
pin-curls 75, 78, 81
plaits and plaiting 16, 17, 42, 74, 75, 101, 137–9
platinum rinse 110
pony tails 42, 65, 132
posture 59, 71

powder shampoo 23
pregnancy 19, 32, 40
prices 23, 53, 61, 102, 103, 110, 111, 117, 118, 122, 132, 140
protein 3, 22, 39, 40
psoriasis 41
puberty 6, 19, 32, 42, 133–4
punk styles 73, 85, 124

quiff roll 136

racial grouping of hair types 6, 10, 14–18
rags 77, 137
razor cutting 72, 73
redheads 6, 10, 109
regrowth of coloured hair 111, 113–21, 124–8
restructurants 32–3, 35, 47, 115
retouching 114, 118, 121, 124, 127, 128
rhubarb root 122
rik-rak winding 97
rim lights 116, 118–19
rinsing 23, 25, 26
rinses, colouring 109–111
rollers 11, 14, 37, 42, 43, 62, 63, 75–8
rub drying 91

saffron 122
sage 123
salon, choice of 51–4
salt water 38, 47, 49, 50
Sassoon, Vidal 94
saunas 38
scalp: damage to 17; dry 37; greasy 36–7; over-stimulation 36; condition of for treatments 103, 112. *See also* allergy; massage; patch test; psoriasis; trichology
Scandinavian type hair 11, 15, 36
scissors 44, 72
scrunch drying 91
scurf 36, 41, 130
sebaceous glands and sebum 6,

Index

sebaceous glands and sebum—*contd.*
   14, 28, 29, 34, 36, 37, 89
semi-permanent colour rinses   110–11, 113, 126, 128
setting   32, 63, 67, 75–82, 138. *See also* lotions
'shades' and 'tones'   109, 113, 114
shampoos and shampooing   21–9; for children   130–1; for dry hair   28, 38; for fine hair   11; for greasy hair   33; after swimming   49; after perming   103–4; colour rinses in   110
shimmers   116, 119–20
shine   5, 6, 24, 31–3, 36
shock and stress, effects of   6, 36, 41
short cuts   16, 62–70, 72, 73, 88
silver foil. *See* foil
smoke fumes   38, 109, 121
soap   23
South Sea Islanders   15
spectrum   129
spiral winding   97
splitting. *See* ends, damage to
spots   28
sprays, colour   125
stack perm   16, 97
static electricity   12, 32, 36, 37, 90
straight hair   5, 10–12, 14, 62, 68, 74
straightening   15, 17, 18, 43, 90, 105–6
streaks. *See* highlights
strength of hair   5, 11, 13, 35, 36. *See also* breakage; damage; ends
student hairdressers   102
style, choice of   55–7, 60–2. *See also* body shapes; clothes; colour; cutting; face structure; fringe; hair type; long hair; make-up; short cuts
sun, effects of   7, 38, 47, 49, 50, 110, 116, 122

tangles   24–5, 27, 32, 37, 131, 138
texture of hair   5, 9–18. *See also* coarse; fine; medium; normal; perms, effect on structure
thick hair   10–12, 15, 62
thin hair   12, 17, 19, 62. *See also* baldness; loss of hair
tinted lights   116, 118
tinting   38, 42, 103, 108–14
tips, colouring of   108, 116, 119
'tones' and 'shades'   109, 113
tongs   14, 37, 42–4, 63, 91
top knot   139
tortoiseshell lights   116, 119
towel drying   83–4
trichology   3, 7, 38, 41, 42

vegetable colours   110–12, 120–4
virgin hair   95, 110
vitamins   39, 40
volume   63–8, 70, 88, 90

walnuts   123
washing hair   24–9
water: elasticity of hair in   8; combing of wet hair   27. *See also* moisture
water, chlorinated   47, 49, 50
water, hard and soft   26–8
water, salt   38, 47, 49, 50
waves, setting and styling of   67, 68, 76, 78, 79, 81
waviness   11, 15
weave winding   100, 101
wedding hairdressing   141–3
white hair   6, 122
widow's peak   18
winding for perms   93–5, 97, 98, 100–2